THE SOCIAL CONTEXT OF
MATHEMATICS EDUCATION:
THEORY AND PRACTICE

*Proceedings of the Group for Research into Social
Perspectives of Mathematics Education*

M NICKSON and S LERMAN (Eds)

SOUTHBANK
press

ISBN 1 874418 03 9

South Bank Press
103 Borough Road
London
SE1 0AA
United Kingdom

British Library Cataloguing-in-Publication Data. A catalogue record for this book is available from the British Library.

"Mathophobia, Pythagoras and Roller-Skating" by Richard Winter appeared in *Science as Culture, No. 10, 1991* and "PGCE Students and Investigational Approaches in Secondary Mathematics" by Barry Cooper appeared in *Research Papers in Education Vol. 5 No. 2*. Both items are reproduced with permission of the publishers.

Origination by Roger Burnett (MRO), Learning Resources, South Bank University.
Printed and bound by Ashford Press, Southampton.

CONTENTS

Preface : by Professor Stieg Mellin-Olsen

Foreword : by Stephen Lerman and Marilyn Nickson

PREFACE

The idea that mathematical knowledge, as other knowledge, is created by human beings whose thinking is influenced by an historical and political context still causes emotional reactions among many. These reactions can be especially negative when someone hints that social factors affect the development of mathematical knowledge and, in turn, the conditions for learning mathematics.

Some educators may say things like, "Don't play around with mathematics. All pupils should receive mathematical knowledge as an ingenious gift of present and former cultures, and we should learn to appreciate its power in and for society." These educators may talk about society's needs with respect to mathematics in such a way that suggests that pupils themselves are not a creative and operational part of that society. 'For society' usually does not mean 'for the pupil as a creative member of society' or 'for pupils as active participants in a society'. When someone tentatively expresses the view that perhaps pupils should be encouraged to use mathematics in creative ways in the social contexts in which they live, even liberal educators might raise their eyebrows and counter this with "Clearly, we already explore uses of mathematics in the school environment, but pupils' creative use of mathematics in their environment - what does this imply?"

We seem to focus on only one end of a spectrum of the potential functions of mathematics education. This is the end of the spectrum where mathematical knowledge is seen mainly as the gateway to higher education. At the other end, the function of mathematical knowledge can be seen as acting as a tool for individuals to participate creatively in the society in which they live, and of which they are a part. "Impossible" representatives of the old school of mathematics educators would say. "Even if we sympathise with this view you well know there are exams, parents, governors, all outside pressures on schools and teachers. Such a view is a luxury we cannot afford." We now see the effects of ignoring this view.

How can a science, such as the didactics of mathematics, progress or even survive, if it does not systematically and persistently challenge and transcend its former discourse, i.e. the structure of the relationships which are fundamental to its basis as a science? Such scientific work is difficult. Those who attempt it rarely receive accolades from those controlling the prevalent scientific discourse.

For this reason it is a special pleasure to introduce this book on research into the social context of mathematics education. I have tried, briefly, to describe the scientific context in which I interpret the work presented here. Authors appearing in the book are members of a group who are part of an ever-increasing band from all over the world who are moving mathematics education onto more fertile and, hopefully, more fruitful ground.

Professor Stieg Mellin-Olsen
Bergen, Norway.

Foreword

This is the first volume of the proceedings of meetings held by the Group for Research into Social Perspectives of Mathematics Education, between 1986 and 1989. The papers represent views of some of the more recent fundamental change in thinking in relation to mathematics education. Much of this thought is an attempt to bring mathematics more firmly into the arena of everyday life and to present the discipline as one which is powerful in the way that it informs decision-making, and helps people to interpret and order their world. This is an intention which, on the surface, may not appear different to what mathematics education has always set out to do. However, mathematics educators are the first to admit a lack of success in achieving this aim for too large a proportion of those we teach. This has led many to re-examine beliefs about the nature of mathematics and to question long established epistemological views of the discipline and to probe for ways in which to make the subject more accessible to more people.

Traditionally, mathematics has been seen as a form of knowledge unlike any other. The formalist school of thought in particular has emphasised this difference between mathematics and other kinds of knowledge, holding that it consists of unchallengeable, abstract truths whose foundations lie far removed from human activity. This emphasis upon the erudite and abstract has served to alienate most people from the subject, and from achieving an understanding of it and an appreciation of its power. More recently, however, this epistemological view of mathematics has been challenged and the results of such questioning have led to alternative interpretations of the foundations of mathematical thinking. A major influence with respect to this change in focus has been that of Lakatos and his work, together with that of other philosophers such as Popper and Kuhn, and it has provided an important stimulus to mathematics educators to explore the implications of these new lines of thought for their discipline. More important than merely acting as a stimulus perhaps, is the fact that it has also given many the courage that comes with a sound theoretical rationale, to adopt different approaches to mathematical teaching and learning.

These new interpretations may best be characterised as social in nature since the emphasis has changed from mathematics as 'given' and beyond question, to mathematics as founded in shared human activity, with a history of development that has a social context just as is the case with any other area of human knowledge. This leads to the opening up of the subject and a consideration of it in terms of theories and problems that are shared, tested, solved or reformulated, and in which anyone from the scholar to the pupil in the classroom can engage. A view of mathematics from this social perspective thus provides a background for potentially powerful and effective developments in mathematics education.

Although research and investigation at these foundational levels in mathematics education are a phenomenon of recent years, there has been an awareness for some time that one's view of the nature of mathematical knowledge, of the function and purpose of mathematics education, and of the learning process, are significant factors in what mathematics is taught, and how. Many of the authors of the papers in this volume have been working in this area for some years, and references have been made in the papers to work going on elsewhere in the world. One can mention Stephen Brown, Thomas Cooney and Jere Confrey in the United States, Hans-Georg Steiner in Germany, George Booker in Australia and Stieg Mellin Olsen in Norway, as amongst those researchers in other countries. There is also an International Group for the Theory of Mathematics Education (TME) which was established at the Fifth International Congress on Mathematical Education in Adelaide in 1984, and has met as a working group at most of the annual gatherings of the International Group for the Psychology of Mathematics Education since then, as well as holding its own conferences.

This leads to the question of why these kinds of investigations are occurring now, or perhaps why they did not occur before. Conditions in different countries, in relation to education in general and to mathematics education in particular, are of course extremely varied, and hence a brief speculation on possible explanations regarding just Britain might be appropriate here.

Until comprehensive education, certainly the major, and perhaps the sole influence, on the mathematics curriculum in this country came from the universities. Grammar school mathematics teachers were university educated, had absorbed the ethos of the university departments, and geared their teaching in this direction. Public examinations also directed school mathematics towards providing the future university mathematicians and scientists with the required knowledge. Any other purposes for learning mathematics were incidental, and such knowledge was expected to be acquired very early on in schooling. Even the New Maths was initially intended to improve the flow of people into university mathematics, in order to increase the number of mathematics teachers.

With comprehensive schooling, and the introduction of the CSE examination, mathematics education came to be seen as essential for all pupils, whatever their future occupation might be. This led to many developments in the mathematics curriculum, with their implicit assumptions about why teach mathematics, what to teach, when and also how. Few attempts were made to make these assumptions explicit. Research in mathematics education in universities was growing considerably, but did not seem to affect the vast majority of teachers of mathematics. This was probably due in part to the kinds of research going on, but also a conviction on the part of most teachers, that mathematics was such a special case in education. Whilst textbooks could improve, and the teacher's explanations get better, the game was still about imparting knowledge to silent and obedient pupils, which a small number would be able to play fully, another group partly, and the majority not at all. Mathematics was seen to be about certain absolute knowledge, after all.

In recent years however, significant changes have occurred both from within mathematics and from outside. The influence of Polya and others led to the introduction of problem-solving and investigations. CSE and GCE still only suited

60% of children, yet we were teaching 100% of children mathematics. The supply of mathematics teachers continued to decrease steadily, partly due to low status and low pay, compared to other options for trained mathematicians. School mathematics, in the popular image at least, was seen to be failing to produce numerate people and especially with the skills required by industry and commerce. In the universities, polytechnics and colleges of education there were significant changes too. Previously, whilst most research had been carried out by psychologists who had moved into mathematics education, departments of education began to include people who had been teachers of mathematics, some with backgrounds only in mathematics, and some with mathematics and an interest in sociology or even philosophy of mathematics.

These factors, and others, have led to a recognition that the previously held absolutist view of mathematics education was inadequate, and to an increased interest in revealing the fundamental assumptions and implicit influences that underlie mathematics education. An examination of the plenary papers and research reports of the various international and national groups of recent years shows the increasing importance and centrality of these ideas.

These studies of a fundamental nature are needed more than ever. In Britain, legislative changes have been, and continue to be, frequent and dramatic. They include coursework, GCSE examinations, the National Curriculum, standardised testing at 7, 11, 14 and 16, teacher appraisal and review of 'A' level programmes. We will need to draw upon the sociological, historical, philosophical, political, linguistic and psychological insights of papers such as those that appear in this volume and those to follow, so as to guide school mathematics through the morass of changes in order to achieve major improvements for all pupils.

The papers in this volume have been presented at the conferences of the Group for Research into Social Perspectives of Mathematics Education, between 1986 and 1989. They are grouped into four sections:

(1)   The National Curriculum in Mathematics: A Critical Perspective
(2)   The Social Context of Mathematics Education
(3)   Theoretical Frameworks and Current Issues
(4)   Political Action Through Mathematics Education.

Each section is preceeded by an introduction, which sets out the field of investigation. Many of the papers could appear in other sections in the collection, and one or two have appeared elsewhere but are included here for completeness. The date on which each paper was presented is given in the list of contents.

It is hoped that this volume will serve a number of purposes. First, there is the need to bring to a wider audience these critical studies of ideas and developments in the teaching of mathematics, past present and future. Second, these studies draw on theories. There is a tendency to view mathematics as self-sufficient and self-justificatory and these studies challenge that view. Third, bringing the ideas contained in the papers in this volume to the mathematics education community offers the greater possibility of disagreement, discussion, debate and consequently development of ideas. It is the unanimous view of the members of the group that knowledge grows through criticism, and we therefore welcome the responses that we hope this volume will elicit.

Part 1

THE NATIONAL CURRICULUM IN MATHEMATICS: A CRITICAL PERSPECTIVE

INTRODUCTION

The legislation leading to the establishment of the National Curriculum has given rise to much discussion and concern in the world of mathematics education. The debate has taken place at a variety of levels and continues to do so, from the perspective of the individual teacher to that of the society as a whole, and from the practical level of Attainment Targets to the questioning of the National Curriculum against current theory in mathematics education. The papers in this section reflect concerns at these levels and examine the implications of the legislated context that has been imposed.

Rosalinde Scott-Hodgetts sets the scene by exploring the implications of the National Curriculum for the sociology of the mathematics classroom. She begins by examining a selection of observations made by teachers about their concerns with respect to the National Curriculum and the doubts that emerge about their ability to continue with an investigative approach in their teaching in the course of pursuing Attainment Targets. The likely impact of such change on classroom interaction is identified and the inequality implicit in testing and the labelling that results from it are discussed.

In the second paper, Zelda Isaacson identifies existing contradictions and commonalities within mathematics, in particular the conflict between mathematics as a utilitarian subject and mathematics as a creative subject. She examines the National Curriculum documentation for mathematics and attempts to determine the position adopted with respect to these two aspects of the subject. She concludes that it is important that the aesthetic side of the subject should not be ignored and proposes the possibility that the two mathematics subjects should appear in the curriculum which reflect the dual nature of the discipline.

In the third paper, Paul Ernest examines the aims and philosophy of the National Curriculum in mathematics. He discusses the aims of the social groups which historically have influenced education in the past and links them with particular philosophies of mathematics. In considering these against the background of the National Curriculum, he identifies the overt conflict and exercise of power that accompanied the development of mathematics within it. He concludes that the differing ideologies of the groups which informed this development are not as contradictory as may first appear and that ultimately they are all concerned with preserving similar social interests.

Finally in this section, Richard Noss briefly examines the political perspectives and implications of mathematics in the National Curriculum. He points out several contradictory positions within the documentation and the difficulty in imposing any coherence upon it as a result. Noss proposes the thesis that the National Curriculum is not about content, rather it is about testing. He concludes that the form of the National Curriculum ultimately distorts the teaching and learning of mathematics.

THE NATIONAL CURRICULUM: IMPLICATIONS FOR THE SOCIOLOGY OF MATHEMATICS CLASSROOMS

Rosalinde Scott-Hodgetts

INTRODUCTION

"It's the testing that will bother everybody. I think that the teaching and learning of the subject will be heavily influenced. It is going to be very restricting. What I mean is that even now, without the National Curriculum, when there is any sort of assessment teachers hate their pupils not doing very well. Testing of any kind, at any stage - especially if schools and teachers are going to be judged - increases such anxieties."

The above statement, made by a secondary school mathematics teachers, is, in my experience, a typical first response to a question probing opinions about the introduction of the National Curriculum. Typical, that is, in the sense that the dominating feature, for the teacher (as for the Secretary of State?), is the testing. The effects of the anxiety surrounding assessment procedures will, in my opinion, override all other factors in influencing change in classroom climate. It is my contention that the overall effect of the implementation of the Secretary of State's proposals will be negative, both in terms of pupils attitudes and their levels of mathematical understanding; this despite the fact that they may well achieve higher levels of performance - levels defined by and measured against the proposed attainment targets.

TEACHER INTERPRETATIONS

"I would hate my classroom atmosphere to change. I do not want the 'fun' part of my work to disappear. I would like to rebel against all this. But if my teaching is to be judged using the results of these assessments I guess I will have to conform. Invariably both the success and failure of pupils is directly attributed to teacher performance."

Many teachers believe, rightly or wrongly, that a primary motivation for the introduction of the National Curriculum is a desire to further challenge their competence and value to society. It also seems clear that the popular press are determined to aid this practice, and to fuel dissatisfaction amongst society in general, and parents of the current school population in particular. Partly because teachers believe that they, as well as their pupils, are to be tested and judged in ways that they feel inappropriate, many feel that there is now no option but to "play the game", and it comes across very powerfully that the rules of 'the game' are clearly understood...

Playing the game means:

"...cramming a child for the tests - he will then quickly forget most of his learning."

"...introducing a narrowness of curriculum - less interest, less chance to diversify (more boredom!)."

"Robot production. Where does ingenuity come from? Thinkers? Inventors? ...The National Curriculum is not geared to producing an academically oriented society. We will become a race of 'plebs'."

"Work will be test oriented ... students will have less scope to use their initiative and to think for themselves."

"Project work will also be limited because it is time-consuming."

"I'll be very strongly aware of the attainment targets which the average pupil in my class should be expected to attain, and I'll make sure that they get there, but I know that will be at the expense of other areas of content, or development of problem-solving processes, which I feel are valuable, and may be more relevant to an individual at any moment in time."

"Real examples are encouraged, but even simple situations in real life can lead to some very sophisticated mathematical activity. They (the pupils) are free now to pursue these if the motivation is strong enough. Now, we are going to be much more constraining - to apply too much structure. It's going to be boring, I think."

"The attainment targets will be taken as the syllabus. The excitement and creative aspects will be lost, and instead pupils will be trained to reproduce standard responses to different 'types' of questions in a traditional 'rote' way to the syllabus."

For me, the most poignant statement was made by a primary school teacher:

> "We will lose the freedom to be spontaneous. For example, suppose a child brought in, say, a butterfly, and I could see that it caught the children's imagination. At the moment I could use the magic and wonder evoked to motivate a range of exciting activities. In future we won't be able to do that, but will have to stick rigidly to the syllabus."

It could be argued that there is nothing in the proposals that is aimed at stifling creativity and spontaneity. Indeed, the Report of the Mathematics Working Group, commissioned by the Secretary of State to inform his decisions lays great stress on the need to build upon current good practice. However, all the teachers quoted in this paper have read the documentation, have decided upon an interpretation of intent - have decided, if you like, which are the 'important bits', and what the implications are for their future practice. They have concluded with remarkable unanimity that they must go back to a traditional approach, if they are to "do justice to" their pupils and themselves; they simultaneously believe that it is tragic that this should be the case:

> "Pre-Cockcroft, we used knowledge about cognitive psychology and diagnostic testing to probe children's understanding. Cockcroft, Prime and other research has been taken into account in the development of programmes of work currently found in schools. Now it will be wasted. We are taking a step backwards. Who knows how far back? This is despite the reflection about good practice apparent in the Mathematics Work Group Report. The thought and work which have gone into the preparation of the Report are admirable, but because of the assessment procedures many will now give pupils examples similar to items in the test banks, and teach them the skills involved, They'll learn how to play the game, how to succeed. The theory relating to good practice, expounded in the document, will be wasted."

It seems that the teachers I have spoken to feel duty-bound to adopt a radically different role when implementing the National Curriculum:

> "If I am teaching the National Curriculum I'll definitely be less flamboyant. At the moment I have a laugh with my pupils, I think it is helpful and improves attitudes. But now there will not be time."

'There will be less room for exploration and experi-
mentation, because of the pressures of time and increased pressure from Heads of Department and Heads
of School."

'We have put so much effort into making mathematics
more interesting more exciting. We're going to go
backwards. It's not going to be learning of the subject,
it's going to be teaching of the subject isn't it?"

This last quote encapsulates for me the nature of the change perceived as
necessary by many teachers. They feel that they must take back responsibility for
learning from the pupils. They must constrain, transmit, direct. They must be
sombre, be economical of time - they must get their pupils through the hoops.

Suggestions that an investigation approach is not appropriate for achievement
of the prescribed attainment targets could be easily explained if they came from die-
hard behaviourists, who were convinced *per se* that the transmission model of
teaching and learning was the most powerful - especially in mathematics, where
what is to be conveyed is a static body of knowledge (see, for example, Lerman,
1983). However, that is not the situation in this case: the teachers interviewed have,
without exception, a tradition (in some cases, admittedly, only recently acquired)
of working with pupils in an open, flexible way, with a strong emphasis on
investigation and problem-solving. They believe that it is such an approach which
leads to understanding and mathematising rather than simply 'knowing some
mathematics'. They also believe that by emphasising the development of problem-
solving/investigative skills and strategies, they are enabling students to approach
the study of new mathematical contexts with high levels of confidence and
competence.

The problem, then, is not that teachers doubt the feasibility and desirability of
reaching the attainment targets by creating stimulating learning environments, in
which their students can explore the various concepts. Rather, the problem lies in
the **framework** of the proposed assessment procedures. The frequency of the
testing, the emphasis on being able to demonstrate particular skills and knowledge
acquisition at particular points in time, and the wide-ranging nature of the content
areas to be assessed at each level, have definite implications in terms of planning
programmes of work. The message to teachers is quite clear. They feel, in essence,
that message is as follows:

1.   There is no value attached to the development of those generalised, transfer-
       rable skills and strategies which in the long term lead to greater mathemati-
       cal maturity, and the ability to cope well with novel situations.

2.   Once particular skills, or items of acquired knowledge, have been ad-
       equately demonstrated, it does not matter that they have been forgotten.
       That is, instrumental learning is sufficient for success.

3.   The primary (exclusive?) responsibility of teachers is to ensure that each student jumps through as many assessment hoops as possible. Both the teacher's and student's ability/value to society are likely to be judged using this criterion above all others.

IMPLICATIONS OF TEACHER'S PERCEPTIONS FOR THE SOCIOLOGY OF THE CLASSROOM

> "Much of what we call teaching quality (or its absence) actually results from processes of a social nature, from teachers actively interpreting, making sense of, and adjusting to, the demands and requirements their conditions of work place upon them. In this view, what some might judge to be 'poor' teaching quality often results from reasoned and reasonable responses to occupational demands from interpretive presences, not cognitive absences, from strategic strength, not personal weakness."

(Hargreaves, 1988)

One clear implication of the interpretation made by teachers of the new demands placed upon them relates to the nature of classroom interactions. It seems likely that there will be a tendency to move towards much greater teacher control, in terms both of directing activity and making sure that there are few distractions.

Hargreaves (1988) presents evidence from a number of sources which highlights the tendency to adopt transmission styles in response to occupational pressures. Specifically, in relation to pressures caused by examinations, he offers the following quotations:

> "The work attempted in the classroom was often constrained by exclusive emphasis placed on the examination syllabus, on topics thought to be favoured by the examiners and on the acquisition of examination techniques."

(H.M.I., 1979)

And from a study of a large group of Scottish secondary school leavers, who had just completed their 'Highers':

> "The most common single method of study was exercises, worked examples, prose translations (73%), followed by 'having notes dictated to you in class' (60%) ... One may infer that many felt there had be a conflict between studying for interest's sake and studying for examination success."

(Gray, McPherson & Raffe, 1983)

Evaluations of the School Council Integrated Science Project also found that teachers' insistence on teaching from the board, although this contravened the Project's guidelines, was justified by the teachers concerned because of the presence of examinations (Weston, 1979; Olson, 1982).

It may not always, of course, be the case that examinations force teachers into the adoption of transmission techniques. There are other teachers who may welcome the assessment procedures as an excuse for maintaining, or reverting to, transmission styles of teaching, (Hargreaves, 1988), or even welcome them as a way of motivating disinterested pupils, (Sikes, Meason & Woods, 1985).

Hargreaves' conclusion about public examinations in general is as follows:

> "There are clearly a number of ways in which public examinations have a bearing on the maintenance of transmission styles of teaching. Their importance for the teachers is substantial in a work environment where few other adults directly witness the quality of the teacher's work, examination results provide one of the few public and apparently objective indicators of a teacher's competence (Mortimore & Mortimore, 1984). Teachers ignore the importance of these results at their peril. Indeed, since the 1980 Education Act's requirement that all schools publish their examination results, it is not unreasonable to surmise that the influence of examinations upon teaching quality (or its lack) is likely to grow in years to come."

(Hargreaves 1988)

It is significant that the group of teachers quoted in this paper (all experienced secondary mathematics teachers) are only now, as a result of the current initiative, making the move from accepting examinations as a fact of life (and even as a motivator for change in the case of less enlightened colleagues), to seeing assessment procedures as a constraint powerful enough for them to predict significant changes in their own teaching styles.

As a consequence both of a greater adherence to transmission styles of teaching and the implications of the Secretary of State's reluctance to accept the 'assessibility' of skills associated with communication and the ability to work collaboratively, it seems possible that the amount of group work currently found in mathematics classrooms is likely to decrease. The importance of such activity for the development of mathematical understanding is well documented (see, for example, Hoyles, 1985).

Again, the problem stems from the conflict between long-term goals and relational learning (which are not to be tested and valued) and short-term skill and knowledge acquisition/instrumental learning (which are to be tested and valued). We are indeed going to be taking giant steps backwards in terms of benefits to be gained from peergroup interactions and creative individual and collaborative activities.

However, it is not just the range of classroom activities which is likely to be curtailed. If France's experience of centralised curriculum control has any relevance within our national context, it is likely that there will be a fundamental shift and narrowing in perceptions of the professional responsibilities of teachers.

> "At their most extreme... a French teacher's responses can be characterised as 'meeting one's contractual responsibility' and an English teacher's as a 'striving after perfection'." (Broadfoot, Osborn, Gilly & Paillet, 1988)

One of the most significant differences highlighted by Broadfoot et al, is the relatively narrow area of responsibility accepted by French teachers in comparison to English teachers. This is nicely exemplified by contrasting comments.

French Teachers:

> "For me, to be responsible is to do my job as well as I possibly can, to teach according to the curriculum laid down and to try to follow it as closely as possible. To be 'responsible' means refusing to accept carelessness."

> "Making sure that my pupils acquire the knowledge appropriate to the level of the class and doing this with commitment."

> "Doing my duty to make sure that my pupils acquire a certain body of knowledge. I'm obliged to do everything possible to attain this."

English Teachers:

> "Creating an atmosphere whereby children will learn through experience - moral and social norms, physical skills and aspects of health and hygiene, develop enquiring minds and creativity and generally develop, progress and fulfil their potential."

> "I'm responsible to each child in my class, that he/she develops intellectually, socially and emotionally during the time in my class. Also I'm responsible to the children's parents that they are informed of the problems and needs of their child, and have some say in how their child is treated in my class. Also to the rest of the staff that I do nothing to hinder their relationship with each other and the children."

This last statement is interesting in that it indicates that the English teacher currently feels him/herself to be accountable to others. Indeed, Broadfoot et al., found this to be another major difference between English and French teachers, and suggested that the lack of a clear central prescription of content in England has meant an obligation to explain and justify the practices adopted to parents and others. In France, on the other hand, teachers are recognised as having no control over what is taught or even how it is taught, so they have no responsibility for justifying their classroom practices.

This difference is intriguing not least because the British Government's initiatives were in part justified by the need to make teachers accountable. Of course, one difference between the French and British situation is the issue of teacher appraisal: it has been made clear that teachers, not just students, are to be tested. Their worth will be assessed primarily on the basis, I suspect, of their pupils' test results. Does that mean that they will not be expected to fulfil the other roles which they now take for granted? Roles "relating to responsibilities both inside and outside the classroom, encompassing extra-curriculum and sometimes even community activities, all aspects of school relationships, accountability to parents, colleagues and Head and a strong consciousness of the need to justify his or her actions to others". (Broadfoot et al., 1988)

I strongly suspect that what in fact will happen will be to some extent analogous with what has happened to the nurses where, as I understand it, the Government's much-applauded generosity has resulted in the following scenario:

1.   On the basis of criteria related to perceived levels of responsibility, each nurse has been assigned a grade.

2.   Level of pay is determined by the assigned grade.

3.   Nurses who subsequently take on only those responsibilities dictated by their grade are liable to disciplinary action and suspension (this has already happened to nurses 'working to grade' in Bristol).

So we have a system where one set of criteria is used for assigning a measure of worth (pay for nurses, public labelling for teachers), and an incompatible set of expectations made in terms of what the job entails.

The political motivation behind current initiatives will be discussed later. At the moment, it is sufficient to recognise the possibility that even if teachers' professional responsibilities are not formally redefined and restricted, it is likely that the demoralisation of teachers will lead to their being less willing to devote their energies to the wide range of non-contractual duties which they currently perform. There is also likely to be a reduction of competence and efficiency within the classroom, as both are factors closely linked to personal senses of worth and value (Hargreaves, 1988).

At worst, British teachers might change from being child-centred, caring professionals to curriculum-focussed state servants, down-beaten and despairing. We are also likely to see an acceleration of the trend for teachers to seek alternative careers: a materially rewarding option for mathematics specialists.

INEQUALITY

> "... when we administer 'a battery of attainment tests'
> to an immigrant child, we ask him: "How English are
> you?" When his performance replies: "Not very", we
> say: "You are deprived and handicapped and must
> have remedial treatment until we have exorcised all
> traces of the alien spirit from your academic behav-
> iour.""

Although there is an increasing awareness that examinations and other assessments can be socially discriminatory (Bourdieu & Passerson, 1977; Burgess &
Adams, 1980, 1985; Hargreaves, 1982), it is difficult to see how assessment items
(especially so many assessment items) can be designed to overcome bias in respect
of race, gender and class. However, the design of test items is only the tip of the
iceberg in relation to the implications of the changes for inequality in education.

It could be that the proposed assessment framework, and its implications for
programmes of work, might alter the balance of disadvantage. For example, I have
discussed elsewhere the possibility that more girls than boys are inclined to adopt
serialistic learning styles within mathematics classrooms (Scott-Hodgetts, 1986).
Without a doubt, the changes in teaching style now being predicted will favour
serialists (who like to learn in a step-by-step, cautious way, learning distinct
'parcels' of knowledge initially), rather than holists (who prefer to investigate new
contexts in a more adventurous, open way, before filling in the details). So if all
other factors were equal, we might expect that the average performance of girls
relative to boys would improve: if the overall assessments include a high proportion of coursework and avoid multiple choice questions, the 'improvement' in
performance might be more marked (Gmitrek 1984).

However, the position of groups believed to be disadvantaged within society in
general, and within education in particular, is extremely complex. It is, perhaps, too
soon to predict overall effects for various groups, and certainly it is outside the
scope of this paper. I would, nevertheless, like to indicate factors which could be of
crucial importance:

1. **Expectations of Parents**
   Will members of particular groups be more likely to come under increased
   parental pressure to succeed? Alternatively, will members of other groups
   be disadvantaged by their parents' low expectations?

2. **Available Resources**
   Are members of some groups likely to be even more disadvantaged than
   under the current system, by lack of facilities for home-based study?

3. **Teacher Bias**
   Are individual teacher biases going to be a more significant factor when the
   new assessment procedures are implemented?

4.   **Cultural Difficulties**

Whether because of social class or ethnicity, children bring with them to school different languages, experiences, behaviours, attitudes, values and skills. Are these differences going to be accommodated judiciously within the 'new' system, or are some groups likely to be further disadvantaged?

Broadfoot makes the following comments in relation to four themes in UK assessment policy - comprehensivisation of assessment; criterion-referencing; an emphasis on curriculum process; and breadth and differentiation:

> "Given the continued unwillingness on the part of central government to dismantle any of the more traditional aspects of secondary schools - indeed its commitments to strengthening them through increasing parental choice, scholarships and new examinations, the new assessment procedures are likely to reinforce, rather than redress existing educational divisions and their associated inequalities ... perhaps even more important, little will have been done to break down teachers' stereotypes of 'able' and 'less able' pupils which are still strongly related to the latter's anticipated ability to achieve particular public examination targets. Such preconceptions are in turn related to a variety of class, race and gender stereotypes."

(Broadfoot, 1986)

I would suggest that these observations are even more relevant in the light of subsequent developments.

THE NEGATIVE EFFECTS OF LABELLING

> "We're going to have pupils in our classes who have been labelled by testing at an early age, and my judgement is going to be affected: I'm going to think, "He's not very good. She's very good", and I'll treat them accordingly."

The teacher who made this statement was at least aware of his own tendency to be influenced by labels previously applied. There are links here, of course, with the stereotyping mentioned in the last section, but labelling theory (see, for example, Meighan, 1981) has a much wider range of applications.

A fundamental notion within the field of labelling theory is that of the self-fulfilling prophesy, and many studies have demonstrated the power of this effect (see, for example, Good & Brophy, 1972; Nash, 1976). The significance of the proposed assessment procedures is the way in which pupils are overtly and simplistically labelled at an early age.

Of course, under the new system it is not only pupils who are likely to be labelled - in the long term, teachers and schools may well be assigned labels too. This is not new, and it is interesting that during the preparation of this paper I had the opportunity to talk to a group of students and ex-students in Stroud, Gloucestershire, where there still exist two (single sex) grammar schools and two (co-educational) secondary modern schools. In both cases, the students did not believe that the new proposals would make a great deal of difference. However, their reasons for that belief were extremely pertinent to the current discussion. One of the girls from the Grammar School made the following statement:

> "We're already under a great deal of pressure at school to perform well, and we have exams every year, so I don't think it will make much difference ... last time our teacher read the results in reverse order, for all those who got more than 50%. That's the mark expected of us. Then she read the list of names of pupils getting less than 50. She said that they were a disgrace. That they were wasting a place which somebody else could make good use of. Two girls were crying."

In contrast, one of the ex-secondary modern students had this to say:

> "If we'd have had these new tests, I still don't think we'd have tried. Our school was rubbish. Once we were there we thought, what's the point of trying?"

A second added:

> "I guess we knew we were failures. That was what most people thought. Well, most of the boys anyway. We just made the most of it ... I mean we had a laugh. Not the girls though ... they were different."

Interviewer:

> "Different in what way? Why?"

> "I don't know really. They got on, kept working. I don't know why, though, I've never thought about it."

As a result of the new initiatives, we are going to have a hierarchy of success and failure: successful and failing schools employing successful and failing teachers to teach successful and failing students. The associated labelling process, and the self-fulfilling prophesies which result from that labelling bode ill for the future.

POLITICAL INTENTIONS

> "The teacher should never be the servant of the state in the way that he preaches and teaches what he thinks the government would like him to do."

(Baldwin 1926)

The way in which the relationship between teachers and the state has changed throughout 20th century history is a fascinating area for study. One aspect of the changes has been the apparent level of autonomy offered to teachers. This must, however, be viewed with some caution, as deeper analysis tends to suggest that the appearance of autonomy and partnership was in fact part of a hoax, aimed at projecting a particular political image whilst making sure that 'real' power remained at Whitehall.

In the 1920s, for example, central directives were removed, but in a way which did not end central control:

> "It was not sufficient to have cheap teachers but to retain control of them. This control, at a time when the main features of the elementary and secondary school system was already established, was not necessarily highly directive, indeed there were positive advantages in creating the appearance of a decentralised system. The central state was left out of day-to-day running of the system, concentrating on tactical control and provided the ideological arguments about the purpose of education and the role of teachers."

(Lawn & Ozga, 1986)

Even the 1944 Act, often considered to have established power-sharing between government, LEAs and teachers, needs to be regarded in relation to the political ends of the government in power at the time:

> "There may be claims for partnership, for freedom for LEAs and for teacher autonomy, but the legal position remains in line with the centralised assumptions behind the public services in general. There was to be positive central government intervention and it was to be designed to promote equality".

(Shipman, 1984)

In the 1920s, the purpose of manipulating the professionalism of teachers was to curb the movement of teachers to the Labour Movement. In return for the freedom offered, and guaranteed fair treatment of teachers, the state wanted a patriotic (i.e. non-pacifist) teaching force, working with the new investment in 'human capital' and promoting civic harmony. There were hints that self-government might be possible for teachers, in return for public service rather that union loyalty, civil service rather than disruptive servanthood (Lawn, 1983).

In the 1940s, on the other hand, the teachers were assumed to be committed to the restructuring and expansion which was taking place, and although the discussions about partnership focussed on the LEAs, the scale of the task made central direction impossible. Thus it was possible for teachers as well as LEAs to extend their claims to partnership status (Lawn & Ozga, 1986).

What then are the political reasons underlying the present Secretary of State's initiatives? In respect of the relationship between teachers and the state, the significant differences are the overt nature of the intention to control, and the level of detail at which the control is to be applied. It may well be that teachers are correct in interpreting this management style as a reflection of a desire to undermine their position within society. This in turn might be because teachers - as a group of predominantly caring professionals, committed to the ideals of equality of opportunity and individual freedom - pose a political threat to this government. Alternatively, the motivation might be financial: to justify poor pay and conditions for teachers, and to resist claims that improvement of educational opportunity demands improvement of resources, by suggesting instead that all the problems are attributable to teacher incompetence.

I wish to pursue the last of these hypotheses further, and to show that one danger of the implementation of the testing programme associated with the National Curriculum might be that the Government could succeed in "proving" exactly what they want to prove.

The key to an understanding of how this is to be achieved lies in an analysis of the way in which the assessment procedures have been designed, and are to be implemented.

The first claim I make is that it should be easier for children to perform well, as measured by and against the proposed attainment targets. This is because, as mentioned earlier, instrumental learning is likely to be sufficient for success, and there is no necessity for pupils to retain knowledge once it has been assessed.

It also seems possible that, because of their anxiety to do themselves and their pupils 'justice', teachers are likely to teach to the test, possibly at the expense of teaching for understanding.

As a result of these two factors, it is likely that test performance will be relatively good. No doubt, the Government will claim "an improvement in standards" and, further, will attribute that improvement to the introduction of the National Curriculum.

At the same time, the results of various schools will be published, and it is probable that many parents - particularly those of 'advantaged' children - will exercise their choice on the basis of the published results.

Schools where the overall results are relatively poor will be labelled 'bad schools'. It would not be surprising if there were a higher proportion of such bad schools in Labour controlled LEAs. It may be that some of the 'bad schools' are schools that have failed to respond to Mrs Thatcher's advice that children should be taught to add up and multiply, and not to do anti-racist mathematics "whatever that might mean".

The Government will subsequently, perhaps, point out that the 'bad schools' had received precisely the same form of funding as the 'good schools'. It follows that lack of resources could not possibly be the cause of the differences...

CONCLUSIONS

> "We repeat that it is not the intention to advocate a standard curriculum for all secondary schools to the age of 16, not least because that would be educationally naïve. One of the greatest assets of our educational arrangements is the freedom of schools to respond to differing circumstances in their localities and to encourage the enterprise and strength of their teachers."
>
> (H.M.I., 1977)

In this respect I have argued that there is probably nothing naïve about the decision to impose central control over the curriculum. I have also outlined some of the reasons for believing that the results of the current initiative may be severely damaging to the mathematical development of students. I have not suggested solutions to the problems which I predict - I can see no easy solutions. Certainly, we should join schools in responding to the H.M.I.'s call to encourage the enterprise and strength of their teachers. Certainly, we must strive to educate society in general and the parents of our current school population in particular, so that they recognise the dangers. Certainly, we must, as educationalists, and, where appropriate, as parents, demand better provision for future generations of school children. I fear, though, that the extent to which we can minimise the damage will be extremely limited.

REFERENCES

Baldwin, S   (1926)   *"On England"* Penguin, Harmsworth

Broadfoot, P   (1986)   "Assessment Policy and Inequality, the United Kingdom Experience".
British Journal of Sociology of Education 7.2 pp 135-154.

Broadfoot,P.,   (1988)   "What Professional Responsibility Means to
Osborn,M.   Teachers: National Contexts and Classroom
Gilly,M., and   Constraints" in *British Journal of Sociology of Education 8.3*
Paillet, A.

Bourdieu, P. &   (1977)   *"Reproduction in Education, Society and Culture"*
Passerson, J.   Sage, London.

Burgess, T. &   (1980)   *"Outcomes of Education"* Macmillan, Basingstoke
Adams, E.

Cooper   (1979)   "Pygmalion Grows Up: a model for teacher expectation, communication and performance influence".
Review of Educational Research 46, pp 389-410.

Goacher B.   (1983)   *"Records of Achievement at 16+"* York: Longmans

Good T. &   (1972)   "Behavioural Expression of Teacher Attitudes".
Brophy J.   *Journal of Educational Psychology 63.6.*

Gmitrek, R   (1984)   *"Sex Differences in Relation to Different Forms of Assessment in a Mathematics Curriculum Project".*
Unpublished M.A. Dissertation, University of London, Institution of Education.

Gray, J.,   (1983)   *"Reconstructions of Secondary Education"* Routledge &
McPherson,A &   Kegan Paul, London.
Raffe,D.

Hargreaves, A. (1988)   "Teaching Quality: a sociological analysis"
Journal of Curriculum Studies, 20.3. pp 211-231.

Hargreaves,D.	(1982)	*"The Challenge for the Comprehensive School"* Routledge & Kegan Paul, London
H.M.I.	(1977)	*"Curriculum 11-16"*, HMSO, London
H.M.I.	(1979)	*"Aspects of Secondary Education"*, HMSO, London
Hoyles, C.	(1985)	*"What is the point of group discussion in mathematics?"* *Educational Studies In Mathematics* 16.2. pp 205- 214.
Lawn,M.	(1983)	*"Teachers and the Labour Movement 1910-1935"* unpublished Ph.D. Dissertation, Open University.
Lawn,M. & Ozga, J.T.	(1986)	*"Unequal Partners: teachers under indirect rule"*. *British Journal of Sociology of Education* 7.2. pp 155-168.
Lerman, S.	(1983)	*"Problem-Solving or Knowledge Centred: the Influence of Philosophy on Mathematics Teaching"*. *International Journal of Mathematical Education in Science and Technology*, 14, 1, pp 59-66.
Marsh, A.	(1971)	*"In a Second Culture"*, T.E.S. 18th June.
Meighan, R.	(1981)	*"A Sociology of Educating"*, Holt, Rinehart & Winston, London
Mortimore,J. & Mortimore,P	(1984)	*"Secondary School Examinations: The Helpful Servant, Not the Dominating Master"*, *Bedford Way Papers* No 18. University of London Institute of Education.
Nash, P.	(1976)	*"Teacher Expectations and Pupil Learning"*, Routledge & Kegan Paul, London
Olson, J.	(1982)	*"Innovation in the Science Curriculum"*, Croom Helm, London
Scott-Hodgetts, R.	(1986)	*"Girls and Mathematics: the Negative Implications of Success"*. Burton, L. (ed.) *"Girls into Maths Can Go"*, London: Holt, Rinehart & Winston.
Shipman, M.	(1984)	*"Education as a Public Service"*, Harper & Row, London
Sikes, P. & Meason, L. & Woods, P.	(1985)	*"Teacher Careers: Crisis and Continuities"* Lewes: Falmer Press
Weston, P.	(1979)	*"Negotiating the Curriculum"*, NFER-Nelson, Windsor

IS THERE MORE THAN ONE MATH?

or

What does the national curriculum tell us about our philosophy of mathematics education?

Zelda Isaacson

INTRODUCTION AND BACKGROUND

The history of the Thatcher government in relation to education is one of a steady attrition of critical and reflective voices. This is exemplified by the gradual (and sometimes precipitate) erosion of philosophy, both as an academic discipline in its own right, and as a part of teacher education courses - and its substitution with so-called 'professional' courses. More recently, we have experienced the unseemly haste with which the national curriculum was rushed through to the statute books with wholly inadequate time for proper consultation and debate.

As a consequence (I believe of both the above) the national curriculum exhibits a lack of any adequately argued or even implicit coherent underlying philosophy. This point has been excellently made by, for example, White (1988). Within mathematics in particular, as Richard Noss points out in a paper given at a recent Research into Social Perspectives in Mathematics Education conference (Noss, 1989), contradictory viewpoints and stances emerge without the authors of the documents apparently being aware of this. Noss refers to the earlier documentation (DES, 1988a and DES, 1988b) which preceded the statutory instrument (DES, 1989) and the non-statutory guidance (National Curriculum Council, 1989). Here, I shall focus on the latest, that is, 1989 documents, but should emphasise that their appearance in no way invalidates Noss' arguments.

Noss recommends that we use the contradictions and anomalies in the proposals (now legislation) 'to subvert the implicit and explicit intentions of those responsible for their introduction' - intentions which he regards as 'profoundly and intentionally anti-educational'. While I agree with Noss that much of the national curriculum could have - whether intentionally or otherwise - undesirable educational effects in mathematics classrooms and elsewhere, I do not intend to respond to this aspect of his paper directly here.

The aim of this paper is rather to focus on the contradictions and anomalies themselves, examine the form in which these appear in the latest versions and explore the source of these differing (and possibly conflicting) perspectives. I believe that the national curriculum documentation has thrown into stark relief a deep-rooted, as yet unresolved - indeed not yet seriously debated - contradiction at the heart of mathematics education. A thoughtful and open debate on the aim(s) and purpose(s) of mathematics education as we head towards the 21st century - a debate which could and should have happened before the legislation entered the statute books - remains an urgent necessity.

It may seem foolish to begin the debate now, rather than putting all our energy into averting the worst of the damage in which too slavish following of the national curriculum could result. I believe we must do just that - although hopefully not at the expense of damage limitation. Legislation is, after all, not written on tablets of stone, amendments can be made, governments do (eventually) change - and in the meantime there will be further non-statutory guidance from time to time which we could attempt to influence.

As I perceive it, the major contradiction within the national curriculum documentation is that between mathematics as a necessary and utilitarian skill for the market place, and mathematics as an aesthetic subject within which individual creative effort plays a central role. After outlining the structure of the statutory instrument for mathematics I look at what the latest documentation actually says about utility vs aesthetics. Then I argue that the confusion in the documentation mirrors that which exists within the mathematics education community. A recent issue of a mathematics education journal is used to substantiate this claim. Finally, I offer an initial contribution to the debate which this paper, taken as a whole, is urging. This last will be in the nature of a kite-flying exercise which I hope will spark off further debate and discussion.

MATHEMATICS IN THE NATIONAL CURRICULUM: THE STATUTORY INSTRUMENT AND NON- STATUTORY GUIDANCE

The statutory instrument consists of a series of Attainment Targets (AT1 to AT14, covering different aspects of the mathematics curriculum) and so-called Programmes of Study. Each attainment target is sub-divided into levels, within each of which there are a (varying) number of statements of attainment with associated examples. The Programmes of Study consist (almost exactly) of the statements of attainment for a particular level, grouped by topic - Number, Algebra etc - without the examples. Thus there is, for example, a Programme of Study for Level 4, consisting (more or less) of the level 4 statements of attainment extracted from the fourteen Attainment Targets. To make a teacher's life even more complicated, the statutory instrument describes four age related stages - Key Stages 1 to 4. Each Key Stage draws on a range of levels. Thus, for example, Key Stage 3 is the lower secondary years, roughly ages 11 to 14. A pupil at the end of Key Stage 3 might be, according to the statutory instrument, at any of the levels 3 to 8. Given that a pupil entering secondary school at age 11 might be at any of the levels 2 to 6, this means that a teacher of lower secondary pupils has to be familiar with all the statements of attainment across fourteen attainment targets for seven levels.

The non-statutory guidance then attempts the tasks of interpreting and explaining the statutory instrument, and offering practical help towards classroom implementation. This document is helpful, and in general professionally sound. However, it does not have the force of law and it is very likely that less attention will be paid to them than they deserve. Teachers who are sinking under the weight of dozens of statements of attainment, and are trying to make sense of attainment targets, levels, key stages, programmes of study and the interrelationship of all these will not always be able to afford the luxury of studying the non-statutory guidance in detail.

I have written the paragraphs above in the main to make the point that any individual teacher or Head of Department has an extremely complex task when they prepare schemes of work or set about establishing a system of record-keeping to fulfil their statutory obligations. Teachers may reasonably be excused if while attempting to see some trees in these tangled woods they are not always alert to issues such as creativity or equal opportunities - the latter almost totally ignored in the documentation. It is extremely difficult to hold the complexities of the structure of the national curriculum in one's head, marry it up to one's existing provision, think about where changes need to be made if one is not to fall foul of the law **and** still attend to these issues.

So what do these documents say about creativity and utility? Predictably, given the monetarist philosophy of the current government, the utilitarian aspect is emphasised in the statements of attainment (and hence Programmes of Study). However, the preamble to both of Attainment Targets 1 and 9 suggests the possibility of investigation and exploration. AT1 for example, says: "Pupils should use number, algebra and measures in practical tasks, in real-life problems and **to investigate within mathematics itself**" (my emphasis). In the examples offered, also, and in the non-statutory guidance (as well as in the earlier 'consultation' documents) both aspects can be discerned. They do not, however, appear in the same way or with equivalent force in these different places.

Closer examination of AT1, for example, suggests that "investigating within mathematics itself" is to be restricted to rather closed tasks, for example (at level 7) "explore the validity of statements about decimal representations such as: "1/13 has a cycle of six repeating digits". This is an example associated with the following statement of attainment:

"follow a chain of mathematical reasoning, spotting inconsistencies; follow new lines of investigation using alternative approaches". The notion that this kind of activity might be conducive to individual creative effort, or might open up the aesthetic aspects of mathematics is not addressed.

It is interesting to note that the more open-ended example tasks which do appear in AT1 tend to be those related to some sort of design or practical problem solving situation: (level 8) "Decide where to put a telephone box in the locality" and (level 9) "Design a wire frame lampshade with the design showing clearly the length of wire and area of material required". Is creativity, then, primarily to be 'applied' to 'useful' tasks? [It should perhaps be emphasised, also, that AT1 and AT9 are somewhat mavericks within the context of the national curriculum as a whole. They are the remnants salvaged (by mathematics educators) from the otherwise jettisoned (by Mr Baker, the then Secretary of State for Education) Profile Component 3 which sought to recognise the importance of personal qualities and communication skills as well as practical applications of mathematics and give all these due weight in the assessment procedures.]

So one way in which the non-utilitarian aspect appears (in a fairly vague way) is within the attainment targets themselves - but only a few of these. Indeed, it is necessary to hunt through the documentation to find examples of this, and once found, neither the detailed statements of attainment nor the associated examples emphasise this aspect. Further, creativity within a utilitarian context seems to be required, rather than within a 'pure' or primarily aesthetic context.

The second way is when a utilitarian sounding statement of attainment has an aesthetic or creative sounding example associated with it (and these too are rare). For instance, attainment target 10 (shape and space) at level 2 has this statement of attainment: "recognise squares, rectangles, circles ... " while the associated example reads: "Create pictures and patterns using 2-D shapes or 3-D objects". Here, the impression given is that the aesthetic aspect is the poor handmaiden of the 'really important' utilitarian outcome. In other words, the creative aspect seems to be merely a tool for 'delivering' given content in perhaps a more palatable form.

The non-statutory guidance, however, offers, at any rate at first sight, a quite different perspective. There, early paragraphs suggest a parallel and equal importance for both aspects. Section A paragraph 2.1 talks about tackling "a range of practical tasks and real-life problems" while 2.2 says "mathematics also provides the material and means for creating new imaginative worlds to explore". Although, as one reads further in the non-statutory guidance, the utilitarian aspect is again emphasised, the reader's attention is brought back, from time to time, to the value and importance of open-ended tasks and the deficiencies of an approach which (Section D paragraph 2.3): **"fails to provide pupils with insights into the unique character of mathematics, the opportunities it gives for intellectual excitement and an appreciation of the essential creativity of mathematics"** (my emphasis). Sadly, the force of this bold statement is somewhat dissipated in the next sentence when we are told that "this aspect of mathematics" will enable pupils to "utilise *(sic)* the power of mathematics in solving problems...". Wholly lacking is any acknowledgment that these two aspects of mathematics might be in conflict within the classroom. Worse, illustrative examples within the non-statutory guidance, like the examples in the statutory instrument, give the impression, once again, that the creative aspect is very much the poor relation in this mathematical family. This is not made explicit, however, nor is any rationale given for the evident emphasis which appears. The contradictions and anomalies are not even noted, let alone addressed and resolved.

Are we to believe that the creative and utilitarian aspects of mathematics carry equal weight or that the creative aspect is there primarily to help along the 'really important' utilitarian aspect?

What do teachers and educationists believe?

This confusion mirrors that which exists among mathematics teachers and educators. I believe it arises because until quite recently (in terms of the history of mathematics education) school mathematics was seen as purely utilitarian by the vast majority of teachers. Gradually, the belief that quite young children could 'be mathematicians' at their own level, make mathematics their own, and appreciate the aesthetic aspects of mathematics has taken root within the mathematics education community. The growth of groups such as the Association of Teachers of Mathematics is living proof of the strength of this idea. The National Criteria for Mathematics which all GCSE courses have to satisfy also bear witness to this. Aim 2.12 reads "produce and appreciate imaginative and creative work arising from mathematical ideas". In my discussion of GCSE mathematics (Isaacson, 1987) I wrote: "The notion that mathematics for all (rather than mathematics for a gifted few only) can and ought to have an aesthetic component, and feed and help develop our creative and imaginative faculties is much to be welcomed". However, it *is* a relatively new idea. Often, creative work has been 'smuggled' into classrooms through the back door by saying that this work will help pupils develop essential skills. Although individual teachers may have believed in the value **in its own right** of this aspect of mathematics, this has not had time fully to take root. Teachers have protected themselves (and their pupils) from attacks of irrelevance and time wasting by pointing to the **usefulness** of, e.g. pattern spotting in a number investigation.

An examination of almost any recent issue of a mathematics education journal reveals the same lack of consensus about the purpose(s) and appropriate content of school mathematics. In the latest issue of **Mathematics Teaching** (MT128, 1989) there are articles advocating imaginative and visual approaches to mathematics in the classroom (eg by Cyril Isenberg and by Ulrich Grevsmuhl). There is an article on measuring which clearly assumes that measuring is mathematics (but is it? - yes, I do know it is one of the national curriculum attainment targets!). Mary Harris pertinently points out that bodies such as the DTI and the Training Agency are "really powerful" (which they are) and that we haven't, as a mathematics education community, taken on board their, and industry's demands (entirely true). A letter from Heather Scott has a careful look at the different meanings "investigating" has for different people - and points out that 'it (using the word "investigation") does not clarify whether pupils are learning maths or not'. How could it when we don't agree on **what** pupils should be learning when they learn 'mathematics'?

This selection from just one issue of **Mathematics Teaching** illustrates, I think, the point I am making. I believe that it is of critical importance that as a mathematics education community of teachers, teacher educators, academics and researchers, we clarify, through detailed discussion amongst ourselves and with outside bodies, what our aims are in mathematics education and how best these aims can be realised. If there are at root unresolvable conflicts between opposing aims, both (or all) of which are valuable, then we need to look for ways of holding these in a creative tension. We cannot simply ignore these realities.

The science education community has already engaged in this exercise to good effect. Undoubtedly, the national curriculum science documentation is less fragmented than ours. Science educators tackled the problem of reaching a consensus at an earlier stage than ourselves for the central reason that science was not compulsory for all pupils to age 16. They needed to get their house in order or they could have seen their subject fade away. The seriousness of this for the nation's future alerted outside bodies to work with teachers to avert a potential crisis. By the time the National Curriculum came along, the science education community was ready with well worked out views, clear arguments, and a powerful back-up lobby. This is what we now need to work towards.

Where can we go next?

As an initial contribution to the debate, I offer the following. I believe that the confusion and lack of consensus, in part at least, comes from lumping together what are really two subjects under a single umbrella heading. Just as there are two subjects 'English language' and 'literature', and analogously two subjects 'technical drawing and craft skills' and 'fine art', there are two subjects within mathematics.

Mathematics **is** both an arts subjects, with all this implies in terms of having an aesthetic component and being a vehicle for the development of creativity, and a utilitarian subject required for everyday life and as a necessary skill (at varying levels) in the market place. Of course there are overlaps. There always are in linked arts/utilitarian subjects. One cannot write a short story without a reasonable grasp of language, and conversely, writing the story improves one's language skills. Similarly, one cannot paint a scene in the park without some rudimentary craft skills, and, again, the skills practised in art classes are handy when one is painting the window frame.

It is the same with mathematics. Creativity in mathematics depends upon at least a basic level of skill (more, or less, depending on the activity). The activity itself then provides a means of improving the skills involved. We need to be clear, however, whether the creative activity is offered to pupils **because it is good in itself** (as is the case with art and literature), or merely as a more palatable way of getting children to practice necessary, otherwise boring skills.

I believe, very strongly, that we ignore the creative and aesthetic aspects of mathematics at our peril. I also believe that to see these primarily as a way of sugar-coating the pill of essential and useful mathematical skills is to deny the central place which aesthetics ought to have within mathematics education - a place which it deserves in its own right.

Perhaps one way forward is to make a case for **two** subjects on the timetable, say Maths A and Maths B, on the model of English Language and English Literature? This would make the point that Maths B was more than just a means to the end of Maths A. It would also leave the way open for pupils to choose to drop Maths B in the last years of schooling while continuing to develop their utilitarian mathematical skills. (This assumes that all pupils meet Maths B during, say, their primary and lower secondary years). I see no need to insist that everyone enjoys maths for its own sake, any more than that everyone becomes an opera fan. Those opting for Maths B could also be exposed to ideas such as mathematical rigour, proof and elegance.

Whatever choices we make, it is essential that the choices are ours, and not imposed on us (and our pupils) by those less well placed to make them. This implies that we enter the political arena, build up a powerful and well informed lobby, and clarify our thinking through both internal and external debate as a matter of the greatest urgency.

[An earlier (and less full) version of this paper was published under the title **Taking Sides** in the **Times Educational Supplement,** 27.10.1989, p.54]

REFERENCES

DES   (1988a)   *Mathematics for Ages 5 - 16*
Proposals of the Secretary of State for Education and Science and the Secretary of State for Wales, Department of Education and Science and the Welsh Office.

DES   (1988b)   *National Curriculum: Task Group on Assessment and Testing*
A Report, Department of Education and Science and the Welsh Office.

DES   (1989)   *Mathematics in the National Curriculum*
London: HMSO

Isaacson Z   (1987)   *Teaching GCSE Mathematics*
Hodder & Stoughton

Mathematics Teaching 128
(September 1989) Articles by R Smith (pp3-4); C Isenberg (pp 6 - 9); M Harris (pp18 - 19); U Grevsmuhl (pp 29 - 35) and letter by H Scott (p 36).

National Curriculum Council   (1989)   *Mathematics: Non-Statutory Guidance*
The National Curriculum Council

Noss R   (1989)   "The National Curriculum and Mathematics: A case of divide and rule?"
Paper given at *Research into Social Perspectives of Mathematics Education* conference held at the Institute of Education, University of London, 16.6.1989

White J   (1988)   "An Unconstitutional National Curriculum" in:
Lawton D & Chitty C (eds)

"The National Curriculum"
Bedford Way Papers 33; London: Institute of Education
pp 113 - 122.

THE NATIONAL CURRICULUM IN MATHEMATICS: ITS AIMS AND PHILOSOPHY

Paul Ernest

INTRODUCTION

The National Curriculum in Mathematics is one component of a major set of structural and contentual changes in state education in England and Wales, brought about by the British Education Reform Act of 1988, and related legislation. Hence, to be understood, the National Curriculum in Mathematics needs to be located within the overall context of social, political and educational policy in contemporary Britain. One approach to this understanding is to consider the range of interest groups at work in educational policy, as well as their underlying aims and ideologies. In this paper I offer a model comprising five interest groups, each with its own ideological perspective, that have a bearing on the development and formulation of the National Mathematics Curriculum. A crucial feature of each group is its aims for the mathematics curriculum. In brief slogan form, the five sets of aims can be described as: 1) narrow and utilitarian 'back to basics', 2) pragmatic and applied mathematics-centred, 3) pure mathematics-centred, 4) progressive and child-centred, and 5) socially relevant and engaged 'mathematics for all'.

The five interest groups represent an elaboration of the historical analysis of Williams (1961). After sketching these interest groups and their aims for school mathematics, I shall try to show the differential impact of their aims in the publications marking the development of the National Mathematics Curriculum.

THE EDUCATIONAL AIMS OF SOCIAL GROUPS

Williams (1961) distinguishes three groups, the industrial trainers, the old humanists and the public educators, whose ideologies have influenced education both in the past and in the present. He argues that these groups exerted a powerful influence on the foundations of the modern British education system in the nineteenth century. He also emphasizes their continued impact on education system, in modern times, for "these three groups can still be distinguished, although each has in some respects changed." (Williams, quoted in Beck, 1981, page 91). Williams' analysis has been widely quoted and utilized (see, for example, Young, 1971; Meighan, 1986). However, I will argue that Williams' analysis of social groups should be extended, and that this extension allows us to describe the evolution of the National Curriculum in Mathematics.

The first of the groups is that of the industrial trainers, representing merchant, managerial and some professional classes, who share the aim of education as preparation for work in adult life. Thus this grouping, according to Williams, have utilitarian aims for education. In modern times, I wish to argue, that two distinct groups can be seen as descendants of the industrial trainers. The first, more radical group are the 'New Right' grouping, who stress 'back to basics' and training for work as their educational aims. I shall retain the title 'industrial trainers' for this grouping. However there is a more progressive strand of utilitarian thought. I shall term this second group of modern day descendants of the industrial trainers the 'technological pragmatists'. this group represents the less dogmatic majority of modern industrialists, employers, managers and sometimes bureaucrats. In distinguishing between the 'politicos' and the 'bureaucrats' within the DES in the 1980s, Lawton (1984) is offering a distinction analogous to that suggested here. The need to distinguish between the two groups will be justified by the divergence between their aims, as explicated in the next section.

Williams' (1961) analysis includes two further groupings to which I shall add a third, the 'progressive educators'. This group embodies the child-centred tradition of primary schooling. Williams may have intended to subsume these under his 'public educator' grouping, but this seems to embody more radical social aims than those of the liberal, progressive educator, now that the shared aim of these two groups of 'education for all' has been achieved. The progressive tradition in education has been identified by many authors, for example Dearden (1968) and Blenkin and Kelly (1981). Thus it seems justified in adding it to Williams' analysis, especially in view of its significance in mathematics education. Thus the Cockcroft Report (1982) stresses the progressive educators aims for mathematics, in combination with others (notably the technological pragmatists, see Ernest, 1991).

Other authors have offered similar if non-identical analyses of social groups with different aims for education. To take just one example, Cosin (1972) distinguishes four groups, which correspond closely to those discussed. These are the rationalizing/technocratic, elitist/conservative, romantic/individualist, and egalitarian/democratic groupings. Although Cosin does not distinguish all five groups, the two new groups that I have added to those of Williams' are evident in this list (the first and third groups). Overall, the effect of adding two further groups to the three identified by Williams is to modernize and refine his powerful analysis, whilst retaining its strengths.

FIVE SOCIAL GROUPS IN EDUCATION

Below, I indicate briefly the five social groups and their aims for education. Subsequently one aspect of their ideology, notably their philosophies (views of the nature) of mathematics, is treated. Since it turns out that one of these social groups is largely responsible for the National Curriculum (the industrial trainers, representing the 'New Right' in British politics), its ideology is given a more extended treatment. Elsewhere, I give a full account of the groups including both their ideologies and their historical development (Ernest, 1991).[1]

THE INDUSTRIAL TRAINERS

The industrial trainers represent that segment of the merchant classes and industrial managers with a 'petit-bourgeois' ideology, who value the utilitarian aspect of education. Thus the aims of the industrial trainers are utilitarian, concerned with the training of a suitable workforce in basic skills. However there is also a powerful social training dimension to their aims, concerned with "teaching the required social character - habits of regularity, 'self-discipline', obedience and trained effort." (Williams, 1961, pages 161-162). This can be traced back to the 'protestant work ethic' and associated values, which are rather puritanical and purist.

In mathematics education, the industrial trainers are the proponents of 'back-to-basics', drill, rote learning and strictly enforced authoritarian discipline. Their aims encompass the mastery of basic numeracy skills as well as 'training in obedience' and subservience inculcated through drill.

TECHNOLOGICAL PRAGMATISTS

The technological pragmatists embody a pragmatic utilitarian tradition in education, which values practical skills, technological progress and the certification of learning, all as a means to furthering an economic conception of society, without the backwards-looking 'social training in obedience' view of the industrial trainers (Golby, 1982). This group represent the interests of industry, commerce, and public sector employers, a perspective shared by many civil service bureaucrats and applied mathematicians. The group is concerned with the acquisition and development of a broad range of knowledge, skills and personal qualities, notably those that prove efficacious in employment. As the level of industrialization has advanced, so too has this group expected more of education, to provide the greater skills required in employment. Indeed, this group sees social development as the outcome of the advance of industrialization and technology. Currently this group is concerned with such issues as information technology capabilities and skills, communication and problem solving skills, beyond the mastery of basic skills.

In mathematics education this group is concerned with teaching mathematics through its applications (but without questioning the nature of mathematics), practical projects, information technology and computer skills. Thus the emphasis is above all on utilitarian and technological mathematics, understood in a broad sense.

THE OLD HUMANISTS

The old humanists represent the educated and cultured classes, such as the aristocracy and gentry. They value the old humanistic studies, and their product, the 'educated man' (the cultured, well educated person), for its own sake. Thus their educational aim is 'liberal education'. That is the transmission of the cultural heritage, which is made up of pure (as opposed to applied) knowledge in a number of traditional forms.

In mathematics the old humanists are the proponents of pure mathematics for its own sake, valuing logic, rigour and the pure beauty and aesthetics of mathematics. In the large, these are pure mathematicians themselves, or the teachers of mathematics in elite schools. Applied mathematics is looked down upon as second rate and besmirched - merely utilitarian. The aims are mathematics for its own sake, and in particular, high level pure mathematics for the mathematically gifted, who will form the new recruits to the group.

PROGRESSIVE EDUCATORS

The progressive educators are the romantic, liberal reformers, whose slogan in an earlier era was typically 'save the child'. This group value education for the sake of the child, to allow the individual to flower, develop and reach her full potential, typically through creativity and self-expression. They are the modern representatives of the progressive tradition, whose proponents have included Rousseau, Pestalozzi, Froebel, Montessori, Dewey, Piaget and Plowden (Dearden, 1968).

In mathematics education, the progressive educators are associated with child-centred teaching, emphasizing active learning, creativity and self-expression through mathematics. This group endorse discovery learning, problem-solving and investigation. They also attempt to protect the child from conflict or failure, by such practices, for example, as never marking their answers wrong (although, as I shall argue, this group still believes in the absoluteness of mathematical truth). Many of these aims are endorsed in the Cockcroft report (1982).

THE PUBLIC EDUCATORS

The public educators represent a radical reforming tradition, concerned with democracy and social equity. Their aim is 'education for all', to empower the working classes to participate in the democratic institutions of society, and to share more fully in the prosperity of modern industrial society. Williams argues that this sector has been successful in securing the extension of education to all in modern British (and Western) society, as a right, through an alliance with others, especially the industrial trainers.

The public educators in mathematics represent radical reformers who see mathematics as a means to empower students: mathematics is to give them the confidence to pose problems, initiate investigations and autonomous projects; to critically examine and question the use of mathematics and statistics in our increasingly mathematized society, combating the mathematical mystification prevalent in the treatment of social and political issues. The outcome should be individuals who are more able to take control of their lives, more able to fully participate in the economic life and democratic decision-making in modern society, and ultimately, able to facilitate social change to a more just society.

Such a view, although marginal in terms of implemented curricula, is embodied in many of the papers in Keitel et al. (1989), and Noss et. al. (1990). A well developed exposition of this view is also given in Abraham and Bibby (1988).

THE PHILOSOPHIES OF MATHEMATICS OF THE GROUPS

One component of the ideologies of the five groups is worth making explicit. This is the range of different underlying philosophies of mathematics. For this shows that despite having very different aims for education, three of the groups share a number of important assumptions. This fact will be drawn on later in an attempt to explain an alliance between three disparate groups with seemingly contradictory aims.

Three broad categories of philosophies of mathematics can be identified: absolutist, progressive absolutist and conceptual change or social constructivist philosophies of mathematics (Confrey, 1981; Ernest, 1991).

Absolutist philosophies of mathematics regard mathematics as a body of fixed and certain, objective knowledge. The traditional philosophies of mathematics such as logicism, formalism and platonism are largely absolutist (Benecerraf and Putnam, 1964). Beyond this, much of general epistemological thought is absolutist in its treatment of mathematical knowledge (see, for example Chisholm, 1966; Woozley, 1949). Most such absolutist views of mathematics are foundationist, regarding the truths of mathematics as being based on certain logical foundations. One consequence is that the structure of mathematical knowledge is seen to be hierarchical, building upwards from its logical foundations by chains of logic and definition. Another consequence is that mathematics is seen as objective, neutral, culture- and value-free.

The ideologies of the industrial trainers, the old humanists and the technological pragmatists each include absolutist philosophies of mathematics (Ernest, 1991). Each of these perspectives views mathematics as a body of certain knowledge, the cornerstone of human knowledge. Each regards mathematics as objective, neutral and certainly value-free, and sees applications as secondary, and not as intrinsic to the nature of mathematics. Each is also happy to accept fixed hierarchical representations of mathematics, such as in the National Curriculum, as representing the discipline faithfully.

The progressive absolutist position is one which also views mathematics as made up of a body of certain, objective knowledge. But in addition, it also accepts that new knowledge is continually being created and added to this body by human creative activity. Confrey (1981) distinguishes this view, which underpins Popper's (1979) epistemology. It also describes the intuitionist philosophy of mathematics, which places human activity as central in the creation of mathematics, and which argues that its logical foundations are never complete (Heyting, 1956). Thus a key feature of this view of this position is that it emphasizes the human processes of knowledge-getting in mathematics, as much as their product, notably, mathematical knowledge.

Progressive absolutism can be identified as the philosophy of mathematics implicit in the progressive educator philosophy. There the certainty of mathematical knowledge is not questioned, but the creative role of human activity in extending it is acknowledged. This is partly why this position emphasises the process and creative human aspects of school mathematics. (It also stems from the ideological model of childhood of this position.)

The conceptual-change, fallibilist or social constructivist philosophy of mathematics can be discerned at least in part in the writings of Lakatos (1976, 1978), Wittgenstein (1978), Bloor (1976), Davis and Hersh (1983), Tymoczko (1986) and Ernest (1991). Although a relatively recent development in the philosophy of mathematics it parallels developments in the philosophy of science (Kuhn, 1962), the sociology of knowledge (Young, 1971), and other fields of thought. This is a social view of mathematics which for all its rigour sees it at base as fallible and corrigible, the ever-changing product of social human creative activity. Thus this view sees mathematics as socially embedded, culture-bound and irredeemably value-laden. It is the philosophy of mathematics of the public educator, where it is combined with a concern for social justice.

In giving an account of the social interest groups in modern Britain, and in mathematics education in particular, a great deal of simplification is involved. The above interest groups are not necessarily stable over time, neither in social group definition terms, nor in aims, ideology and mission terms. Within a single grouping there is likely to be a family of overlapping ideologies rather than just one, and group members may subscribe to composites including components of several of the ideologies. Such complexity is routinely recognised by sociologists studying phenomena such as the professions and the sociology of science (Bucher and Strauss, 1961; Crane 1972) and must be acknowledged here too.

THE DOMINANT GROUP: THE INDUSTRIAL TRAINERS

The five social groups, and their ideologies, can each be identified in the modern British context. Each has had its impact on the development of the National curriculum in mathematics. However the power of these groups, and the strength of their impact, are far from equal. They vary from the overwhelmingly powerful industrial trainers, via a powerful technological pragmatist-old humanist alliance, through the significant but weak progressive educators, to the completely power-less public educators. Below, I shall document how the impact of these groups on the National Curriculum is commensurate with this attribution of their relative power.

In view of the tremendous power of the industrial trainers, who have dominated the frame and form of the curriculum, as well as the other changes affecting the whole education system as a result of the 1988 Education Reform Act, their ideology is worthy of a more detailed analysis. This will also allow subsequently a more detailed and convincing analysis of this group on the National Curriculum in mathematics.

The 'New Right' as the modern industrial trainers

In the Victorian era the industrial trainers succeeded in defining

> "education in terms of future adult work, with the
> parallel clause of teaching the required social charac-
> ter - habits of regularity, 'self-discipline', obedience
> and trained effort."
>
> Williams (1973, pages 161-162)

I want to argue that this describes the position adopted by the 'New Right', the radical conservative cluster including Margaret Thatcher and like-minded members of recent Conservative governments, as well as the associates of a number of right-wing pressure groups and 'think tanks' such as the Centre for Policy Studies (Gordon, 1989). The central figure, and the source of the power of this group is of course the British Prime Minister from 1979 to 1990 and beyond, Margaret Thatcher. This is not only because of her position as Prime Minister, but because of the exceptional powers she has arrogated to this position, and indeed to successive Conservative governments under her leadership. Consequently, the ideology of the New Right is of great importance as the engine of government policy since 1979, across the whole range of social and political issues. Some of the key aspects of this ideology are as follows.

The New Right see tradition and authority as the sources of both knowledge and moral values. Particular emphasis is given to Victorian values and the protestant work ethic, which place a premium on work, industry, thrift, discipline, duty, self-denial and self-help (Himmelfarb, 1987). These are all seen as morally desirable and good. In contrast play, laziness, self-gratification, permissiveness and dependence are all seen as bad. These values are exemplified in the Black Papers (for example, Cox and Boyson, 1975) and many of the publications of the Centre for Policy Studies (for example, Lawlor, 1988).

These values are also associated with a view of humanity and social relations. From the valuing of authority and the traditional Judeo-Christian values comes a hierarchical view of humankind - good at the top (close to God) and base at the bottom. (The pun illustrates how positional language is laden with these values: high and elevated mean good, low and base mean bad.) Such a view sees people as unequal, identifies social position with moral value, sees children as bad, or at least naturally inclined to error (with its moral connotations). Another consequence of this perspective is to trust only those who agree with these views (one of us), and distrust and reject those who disagree (them). The overall intellectual and ethical perspective fits with Perry's (1970) stage of 'Dualism'.

Lawton (1988) distinguishes between two educational philosophies of the New Right. He divides them into the 'privatisers', who want to see all of education privatised and left entirely to unregulated market forces, and the 'minimalists', who want to see a cheap minimal state education system retained. Both strongly agree on a market-place view of education, but differ on how far to take it. However what I describe as the industrial trainers is dominated by the 'minimalists' of the New Right, since strong central regulation and control of education, including the imposition of a basic skills curriculum with "social training in obedience" are deemed necessary by them, as well as adherence to the market- place metaphor (Bash and Coulby, 1989).

Illustrative case study of the New Right: Margaret Thatcher

Margaret Thatcher provides a case study of the viewpoint described. Young's (1989) biography of her is aptly entitled 'One of us', illustrating the importance he attaches to the penultimate point made above.

Thatcher was greatly-influenced by her father, and she adopted his Victorian values of hard work, self-help, rigorous budgeting, the immorality of extravagance, and duty instead of pleasure (Young, 1989, pp.5-6). These values were also reinforced by Methodism (she still reads improving works by moral theologians, **op. cit.** p.409) and said of Wesley "He inculcated the work ethic, and duty. You worked hard, you got on by the results of your own efforts." (**op. cit.** p.420). She has a dualistic view of the world, a belief in right vs wrong, good vs evil, coupled with a certain knowledge in her own absolute and incontestable rightness (**op. cit.** p.216). The us-right, them-wrong view is applied to educationists, who are regarded as incorrect and untrustworthy, as well as to other professional groups. The imposition of 'equalisation' instead of acknowledging ability differences denies "educational opportunities for those prepared to work" (**op. cit.** p.69). She developed a strong market-forces, anti-collectivist and anti- interventionist view, from an early age (**op. cit.** p.82). Thatcher "felt an uncommon need to link politics to a broad, articulated philosophy of life." (**op. cit.** p.405) She believes that the values of free society come from religion, and the key value concerns free will, personal choice, reinforcing the market-forces, anti-collectivist ideology. She said the most important book she had ever read 'A Time for Greatness' by H. Agar, which has the theme that a "moral regeneration of the west [is] needed...we must fight inner weakness."

Elsewhere she has emphasised that she bases on Christianity her (and hence the only permissible) "view of the universe; a proper attitude to work; and principles to shape economic and social life" (quoted in Raban, 1988).

It can be seen that central elements in Thatcher's ideology are Victorian values for the personal (the virtues of work, self and moral striving) and the related marketplace metaphor for the social (industry, welfare and education). The pre-eminence of this social model (in the policies of the government) stems from the adherence of Thatcher and her early advisers K. Joseph and J. Hoskyns, to it. There is also an anti-intellectualism stemming from a dualistic certainty of belief (Warnock, 1989), which sees argument as the attempt to dominate, coupled with a distrust in professionals (they are morally flawed and self-seeking).

Thatcher's moral outlook is in keeping with an ideological model of childhood, implicit in the 'Victorian values' of the elementary school tradition. This sees children as 'fallen angels', who are naughty by nature and must be kept busy ('the devil finds work for idle hands'), and who are 'empty buckets' who must be trained and fed the right facts by the teacher, for left to their own devices their minds will fill up with inappropriate and disorganised material (Ramsden, 1986). This perspective may be identified with the authoritarian 'custodial pupil control ideology' (Abraham and Bibby, 1988, page 9).

From the authoritarian social relations derive both authoritarian teaching and a belief in a rigid ability typology of children. Children are born with different abilities in mathematics, so streaming and selection are necessary to allow children to process at different rates. (The 'better' children are in some sense more worthy, and so to penalize them by 'equalling down' would be unnatural and morally incorrect. The inferior children can better themselves if they try hard enough - through moral self help. The choice of private schools ensures that better children are catered for.)

MATHEMATICAL AIMS & PEDAGOGY

A clear and representative statement of the industrial trainer views on education, including mathematics education, is given in Lawlor (1988). It can also be found in other publications, such as the Black Papers, Campaign for Real Education (1987), Cox and Marks (1982), Froome (1970), Letwin (1988) and Prais (1987, 1987a).

The industrial trainer perspective sees mathematics as a "clear body of knowledge and techniques" (Lawlor, 1988, p.9), made up of true facts and correct skills (as well as "complicated and sophisticated concepts more appropriate to academic research", **op. cit.** p.7) It is clearly demarcated from other areas of knowledge, and must be kept free from the taint of cross-curricular links and social values. (**op. cit.** p.7)

The aim of education is first of all "to ensure that children leave school literate, numerate and with a modicum of scientific knowledge, it should not extend beyond these three core subjects, nor attempt to do more than set minimum standards in basic knowledge and technique." (**op. cit.** p.5)

Thus the overt aim is utilitarian, the acquisition of functional numeracy, although there is also the covert aim of 'practice in obedience' to 'gentle the masses'.

"The acquisition of knowledge require(s) effort and concentration. Unless children are trained to concentrate and and make the effort to master knowledge they will suffer.." (**op. cit.** p.19) "There is no reason to imagine that pupils learn from talking." (**op. cit.** pp.18-19). What is important is paper and pencil work, and drill and rote learning (**op. cit.** p.15). It is wrong to say that learning must take place without effort and in the guise of games, puzzles and activities (**op. cit.** p.7). We should restrict the use of the calculator (**op. cit.** p.15). It is inappropriate to make subject matter relevant to the interests of the child (**op. cit.** p.18)

'Proper teaching' is needed **not** "salesmanship; enthusiastic teaching; surveys of peoples opinions; attractive resource materials; investigative activities; games, puzzles, television material." (**op. cit.** pp.13-14) Teaching is is a matter of passing on a body of knowledge (**op. cit.** p.17).

Clear targets are needed, and children's ability to reproduce knowledge and apply it correctly must be tested (**op. cit.** p.11). Tests provide external standards. If children are protected from failure, the tests are a sham (**op. cit.** p.7). Passing tests correctly is the goal (**op. cit.** p.9).

Educationists oppose these 'correct' views and are therefore untrustworthy (Young, 1989; Lawlor, 1988; Aleksander, 1988). Instead of concentrating on numeracy, they want to tempt children into their own moral slackness, or worse still, into evil, through political indoctrination.

> "children who needed to count and multiply were learning anti-racist mathematics - whatever that might be. Children who needed to be able to express themselves in clear English were being taught political slogans."

> (Thatcher, 1987)

The ideology of the New Right industrial trainer group has been explored in great detail because it plays a central role in determining the National Curriculum, as will be argued, in two ways. First of all, this group fiercely promotes the metaphor of the market place, which underpins much of the government's overall social policy, especially in education. Secondly, the industrial trainer aims and ideology play a central, and through the exercise of power, a dominant role in the development of the National Curriculum in mathematics.

THE GENERAL CONTEXT OF THE EDUCATION REFORM ACT (1988)

Given that the main groups of actors have been identified, the National Curriculum needs to be located, first of all, within its overall political context. We have seen that the New Right clustered around the Prime Minister, have a social philosophy based on the metaphor of the market place, coupled with a moral vision demanding strict authoritarian regulation of the individual. In practice, the Thatcher government has a range of policies concerning industry, commerce and social services based on the market place metaphor (Bash and Coulby, 1989). This is elaborated as follows:

All goods, utilities or services are commodities to be bought and sold in the market place. These commodities are 'manufactured' by 'workers'. The market place is driven by market forces, so that consumer choice is the final arbiter of value, subject only to minimal quality regulation. Both individuals and corporations are solely responsible for themselves, with market forces and competition ensuring that the fittest survive. This metaphor leads to two main currents of policy.

1. The Rolling Back of State Involvement

Maximum freedom of choice and is put into the hands of the consumer. This requires the 'rolling back' of state and local government interference in the market. This is the source of the privatisation of industry and social services and the deregulation of the media, financial markets, etc. It also involves reducing the power of professional groups, who interfere in the workings of the market out of self interest.

2. The Imposition of Centralised Control

This policy seems to be a contradictory to the market-place metaphor, and indeed Maw (1988) argues that it is due to opposing forces in the New Right. However, as the account above show, this can also be explained in terms of the authoritarianism of the New Right. Relating this to the market-place, it can be said that regulations are put into force concerning the mini-mum standards of the commodities on offer. This leads to two approaches to regulation according to the market and the responsibility of the market makers. Private corporations, manufacturing industry, the commercial and financial sectors, entrepreneurs and so on are all responsible market-makers and are allowed to be self-regulating. In general, the professionals involved in social services including teachers, lecturers, scientists, Her Majesty's Inspectorate, doctors, nurses, lawyers, local government officers, and so on, are untrustworthy (they are perceived to hold an incorrect moral outlook) self-interested groups who need strong government regulation to permit market forces to work freely (Aleksander, 1988; May, 1988). (Indeed society needs strong regulating forces, such as the police, to keep the morally tainted populace in check.)

On the basis of these two sets of forces, we can see two sorts of policies at work in education.

1. Consumer choice and power is promoted by encouraging diversity (opting out, City Technology Colleges, discretionary charging and private schools), consumer power (increased parent and local community representation on governing bodies) consumer information (publication of test and exam results). Deregulation is at work in opting out, Local Management of Schools, and in the independence of polytechnics and colleges.

2. Centralised control is imposed differentially, with private education trusted to be self-regulating (subject only to the market). State education is subjected to strict central regulation, to permit market forces to work. This is seen in the imposition of strict conditions of service on teachers, and the imposition of the National Curriculum. The teachers are subject to strictly regulated conditions of service, as befits workers producing goods for the market place, to ensure the delivery of products of at least minimum standards (the 'proletarianisation of teachers', Brown, 1988; Scott-Hodgetts, 1988). The National Curriculum imposes quality control and consumer labelling of educational products, just as with labelled foodstuffs. This ensures that the consumers of education know (i) the ingredients in understood labels (traditional school subjects) (ii) the quantities of the ingredients (time apportioned to subjects), and (iii) the quality or nutritional value of each of the ingredients (national test ratings on the subjects according to TGAT framework). This allows

customers to choose schools or even teachers according to their market value.

What has been described is the commodification of education (Chitty, 1987). Education is a commodity, like any other. Its price is made public in the market, but central government regulation protects consumer interests by quality labelling through the National Curriculum (Pring, 1988). (Of course the imposition of this model of a national curriculum is also determined by the full set of ideological preconceptions of the industrial trainers including their epistemology and educational theories.)

Within this context of 'commodification of education', the development of National Curriculum for mathematics is circumscribed by the imposition of the following severe constraints (Department of Education and Science, 1987):

> 1. rigid, traditional subject boundaries, contrary to modern curriculum thinking and primary school practice (Her Majesty's Inspectorate, 1977);
>
> 2. a single fixed assessment model (the TGAT framework from Department of Education and Science, 1988b) presupposing a unique hierarchical structure of subjects;
>
> 3. an assessment-driven curriculum, requiring the greatest degree of definition for core subjects (mathematics, English and science) in terms of a hierarchy of objectives specified as discrete items of knowledge and skill;
>
> 4. a very short timescale for development and implementation;
>
> 5. severely limited terms of reference for the National Curriculum Working Groups restricting them to formulating clearly specified objectives and programmes of study (subject to all the previous constraints).

Prior to the design of the National Mathematics Curriculum, the industrial trainers have determined the form that the management and organisation of schooling shall take, and have placed severe constraints on the nature of the school curriculum by making sure it is assessment-driven. The only concession is that the actual assessment model is apparently old humanist/technological pragmatist, rather than an industrial trainer-New Right minimum basic skills model. However, this is less of a concession than it might appear. For the hierarchical curriculum framework of the model, which requires that all pupils master each level before progressing to the next, ensures that below-average attaining pupils will go little beyond the basic skills specified for the bottom few levels of the curriculum.

Although the framing of the National Curriculum reforms are due to the industrial trainers, the 'politicos' (Lawton, 1984), the detailed specification of the National Curriculum can be seen as the work of the Department of Education and Science civil servants ('bureaucrats'). This is reflected in the explicitly technological pragmatist aims of the proposals.

> "The aim is to equip every pupil with the knowledge,
> skills, understanding and aptitudes to meet the re-
> sponsibilities of adult life and employment."
>
> Department of Education and Science (1987b, page 35)

In keeping with these aims the reforms give an elevated place to science and technology, for in the National Curriculum

> "The core subjects are English, mathematics and sci-
> ence. The other foundation subjects are technology
> (including design)...The foundation subjects...will
> cover fully the acquisition of certain key cross-curricu-
> lar competences: literacy, numeracy and information
> technology skills."
>
> Department of Education and Science (1989, pages 7-8)

This statement combines the technological pragmatist utilitarian emphasis with that of the industrial trainers on 'basic skills', which are redefined to give greater priority to science, technology and computing. This emphasis is indicated by the prioritizing of mathematics, science and English above other areas of the curriculum both by labelling them as 'core' subjects (meaning central), and by scheduling the development of the curricula in these subjects to precede all others (followed next by Technology).

This represents the background against which the National Mathematics Curriculum was formulated.

THE NATIONAL MATHEMATICS CURRICULUM

In the summer of 1987 a Mathematics Working Group for the national curriculum was constituted. It was made up of 9 mathematics educationists, 3 head teachers, 4 educational administrators, 2 academics, 1 industrialist and 1 member of the New Right, although not all served for the full term. On 21 August 1987 the Secretary of State for Education, K. Baker informed the chair of the group of their report dates (30 November 1987 and 30 June 1988) and task: to design an assessment-driven mathematics curriculum for the age range 5-16 years, specified in terms of discrete items of knowledge and skill. Severe limits were imposed on what the group could discuss, permitting an initial consideration of "objectives and the contribution of mathematics to the overall school curriculum", before focusing on "clearly speci-fied objectives" and a "programme of study" (Department of Education and Science, 1988, pages 93-94).

On 7 September 1987 one of the mathematics advisers in the group circulated a key document including the following statement to the group.

"Global statement
The mathematics curriculum is concerned with:
(a) tactics (facts, skills, concepts)
(b) strategies (experimenting, testing, proving, generalising,...)
(c) pupil morale (pupil work habits, pupil attitudes)
The treatment within the NMC [National Mathematics Curriculum] of what is contained in this statement is, to me, the most important issue facing the working group. There are many possible scenarios, but I will confine myself to two:

Scenario A. The NMC deals relatively thoroughly with mathematical facts, skills and concepts (what I am calling the tactics of mathematics). But then it makes only superficial references to strategies and pupil morale, perhaps devoting, say 5% of the statement to these aspects.

Scenario B. The NMC starts off with a clear statement on pupil morale. This is followed by a detailed statement on general strategies which are the essence of mathematical thinking. Finally, it deals with mathematical tactics. Within this scenario it is strongly emphasised that pupil morale is paramount, followed by mathematical strategies and then mathematical tactics (concepts, skills and facts) - strictly in that decreasing order of importance. In simplistic terms this is based on the obvious principle: forgetting a fact (such as $7 \times 8 = 56$) can be remedied in a few seconds, but bad work habits and poor attitudes are extremely difficult to correct and may, indeed, be irreversible."

Mayhew (1987, p.7)

This is a clear statement of both the mathematics-centred view of the old humanists (and technological pragmatists) (A), and the child-centred progressive educators' view (B), but strongly endorsing the latter. This statement is prescient, in that it sets out the ideological boundaries of the struggle to come. It excludes two views, those of the industrial trainers and the public educators.

The statement begins by assuming the broad definition of the outcomes of mathematics learning identified by Bell **et al.** (1983) and endorsed by Cockcroft (1982), as reformulated by Her Majesty's Inspectorate (1985). This replaces the original notion of 'the appreciation of mathematics', which leaves open the possibility of an awareness of the social role and institution of mathematics, and hence the inclusion of the public educator aims, by 'pupil morale' with its progressive educator connotations.

The initial internal struggle between the old humanists/technological pragmatists and the progressive educationists was apparently won by the latter. The Interim Report (Department of Education and Science, 1987) was a clear statement of the progressive view of mathematics, following scenario B above. Prior to its publication, the sole member of the New Right on the working group (S. Prais) sent a note of dissent to K. Baker, complaining at length that the group were largely "sold on progressive child-centred maths" instead of concentrating on essential basic skills (Prais, 1987, 1987a; Gow, 1988). He resigned shortly afterwards.

Baker dissociated himself from the report by not having his critical letter of acceptance published with it (unlike all the other Interim Reports), and the chair of the working group was replaced. Baker's letter expressed his "disappointment" and re-directed the working group to "deliver age-related targets" with "urgency" and to make "faster progress" and required the new chair to report progress at the end of February 1988 (Department of Education and Science, 1988, pages 99-100).

He also attacked the progressive philosophy of the report and instructed the group to give the greatest priority and emphasis to attainment targets in number. He pointed out "the risks...which calculators in the classroom offer" and stressed the importance of pupil proficiency in computation and the "more traditional paper and pencil practice of important skills and techniques." This attack embodies the back-to-basics view of the industrial trainers, with its emphasis on school 'work' as social training.

On 30 June 1988 the final report of the Mathematics Working Group (Department of Education and Science, 1988) was sent to Baker. Its proposals represent a compromise between the old humanist, technological pragmatist and the progressive educator views of mathematics, as is reflected in the brief discussion of the nature of mathematics (**op. cit.,** pp.3-4). The old humanist and technological pragmatist impact on the proposals is evident from the range, depth and spread of mathematics content in the proposals. Mathematical content makes up two of the three profile components and is given a 60% weighting. The impact of the progressives is shown in the third component comprising mathematical processes and personal qualities, given a 40% weighting. The technological pragmatist influence is particularly evident in the name given to this component: 'Practical applications of mathematics', and in the attention given to technology. The public educator views are nowhere to be seen. Although lip-service is paid to social issues concerning equal opportunities, the need for multi-cultural mathematics is repudiated. Likewise the aims of the industrial trainers are repudiated. For example, the long multiplication and division algorithms are rejected as unnecessary (op. cit., p 9). "There is no 'moral gain' derived from tackling 1000 long divisions when calculators exist" (**op. cit.,** p.8).

Baker accepted the report, presumably because it contained the objectives for assessing mathematics that he required. That is the specification of mathematics content in terms of 12 broad attainment targets each defined at 10 age-related levels (and programmes of study based on this). This represented the old humanist/ technological pragmatist part of the proposals. He rejected the third profile component, particularly personal qualities, which represented the heart of the progressive part of the proposals. He allowed for a token representation of the processes of mathematics, if they could be incorporated with the content targets under a technological pragmatist banner (**op. cit.,** pp.ii-iii).

Baker instructed the National Curriculum Council to prepare draft orders on the basis of these recommendations, and the industrial trainer view that "pencil and paper methods for long division and long multiplication" needed to be included (National Curriculum Council, 1988, p.92). The Council carried out its instructions and published its report in December 1988 (**op. cit.**). The part of the proposal reflecting the progressives' view was marginal, comprising 2 out of 14 attainment targets applying mathematical processes to the content areas defined by the two profile components. Even this concession is phrased in technological pragmatist rather than progressive educator language. The old humanists, however, managed to have an impact on the proposals, through the addition of extra mathematics content at the higher levels of the attainment targets.[2] This impact was achieved during the period of public consultation on the report. Although presenting the appearance of a democratic procedure, this consultation was little more than a sham, for although nearly 80% of respondents argued for the retention of the third profile component, it was dropped as was already indicated in the Secretaries of State's forward to the final report (Department of Education and Science, 1988).

Scenario A (above) has been enacted, representing the triumph of the old humanist and technological pragmatist alliance, with marginal influences of the progressive educators, but within a framework dominated by the industrial trainers. This is reflected in the final form of the National Curriculum in mathematics (Department of Education and Science, 1989). This can be said to embody the aims of three groups. The curriculum represents a course of study of increasing abstraction and complexity, providing a route for future mathematicians, and meeting the old humanist aims. It is a technologically orientated but assessment-driven curriculum, meeting the technological pragmatist aims. It is one component of a market-place approach to schooling, and an assessment-driven hierarchical curriculum with traces of the progressive educators expunged, meeting some of the aims of the industrial trainers. The overall range of content exceeds the basic skills deemed necessary by the industrial trainers. However, the underlying assessment framework ensures that below-average attaining students will study little more than the basics, in keeping with industrial trainer aims. Overall, the outcome is largely one of victory for the industrial trainer aims and interests, together with their allies, despite the progressive climate of professional opinion since Cockcroft (1982).

The overall pattern of shifts in philosophical orientation is illustrated in simplified form in Figure 1. This illustrates how the progressive philosophy of the Interim Report was moved to a compromise with the old humanist/technological pragmatist alliance in the Final Report, as a result of the impact of industrial trainer and other forces. The same forces in collaboration with the old humanist/technological pragmatists resulted in a shift in philosophy (in the Consultation Report, National Curriculum Council, 1988) to a compromise between the old humanist/technological pragmatist position, and that of the industrial trainers. This is the position of the Parliamentary Orders (Department of Education and Science, 1989). If the frame, as well as the content of the National Curriculum in mathematics is considered, then the success of the industrial trainers in dominating the mathematics curriculum must be acknowledged.

Finally, as much as to say that it was an afterthought, the National Curriculum Council (1989) published the non-statutory guidance in mathematics. This embodies some of the progressive emphases stripped from the previous reports (but not

Figure 1

The shift in composite aims of the National Curriculum in mathematics during its development

```
                        AIMS OF THE IDEOLOGICAL GROUPS

              Public        Progressive      O.H./T.P.       Industrial
              Educator      Educator         Alliance        Trainer
                |              |                |               |
Sequence of     |              |                |               |
Documents

INTERIM-------------       X
REPORT                     XXXXX
                           XXXXXXXXXX
FINAL-----------------     XXXXXXXXXXXXXXX
REPORT                        XXXXXXXXXXXXXXXXXX
                                 XXXXXXXXXXXXXXXXXXXX
CONSULTATION--                      XXXXXXXXXXXXXXXXXXXXXX
DOCUMENT                            XXXXXXXXXXXXXXXXXXXXXX
                                    XXXXXXXXXXXXXXXXXXXXXX
PARLIAMENTARY-                      XXXXXXXXXXXXXXXXXXXXXX
ORDERS
```

the interim report, which is expunged from the official history of this development). It makes suggestions as to mathematical pedagogy, which unlike curriculum content and assessment, is not subject to central regulation. Thus, for example, 'open' (multiple answer) rather than 'closed' tasks (unique answer) are recommended. This can be seen to emanate from the progressive sympathies of the education professionals working in the Council, representing a gesture of solidarity with the widespread progressive educator sympathies among teachers and educators. However, it can also be seen as a palliative for the offended sensibilities of professionals in the education service, designed to reassure them that the progressive educator aims are alive and well in the national curriculum. However, the proposals are in fact toothless and marginal, since they have no official standing like that of the curriculum content and assessments. And as we have seen, the real aims of the National Curriculum in mathematics are largely those of the industrial trainers.

COMPARING THE NATIONAL SCIENCE CURRICULUM

It is interesting to apply a similar analysis to the development of the National Curriculum in science. In translating the discussion to the arena of science education, I am making the assumption that the same philosophical positions and groups can be identified with regard to the science curriculum. Whilst in part this is legitimate, it must be said that such positions in science education are of greatly different significance. Thus practical work is traditional in science, and does not have all the connotations of progressive child-centredness it has in mathematics. Thus the industrial trainer position in science does not view practical activity by students as anathema, as it does in mathematics (where it signifies permissiveness). Nor does it see 'social training in obedience' as such an important goal of science teaching as it is in mathematics (and English). The old humanist position applies less well to science, and does not set itself in violent opposition to practical applications of science and technology. For the pure-applied dichotomy does not have the same force in science as in mathematics, where it has strong value connotations, and was traditionally associated with class distinctions ('pure' associated with the leisured and professional classes, 'applied' with the working and mercantile classes). Likewise, appreciating the social context of science is not seen as having the same ideological significance as it does in mathematics, where it may be seen to represent a radical, leftist position. For although, as we have seen, the social- and value-neutrality of mathematics is assumed by all positions other than that of the public educator, this is not the case with science. (Thus the non-statutory guidance in science asserts that "Science is a human construction" National Curriculum Council, 1989a, p.A4, and refers to its origins in other cultures, and the moral and ethical issues its applications raise.)

Despite these differences, it is instructive to plot the progress of the philosophical orientation of the National Science Curriculum, in a similar simplified diagram (see figure 2).

Unlike in mathematics, the Science Working Party had a major public educator strand in their philosophy, as well as progressive and old humanist strands. However, far less was made of the progressive and public educator philosophies in the Interim Report (Department of Education and Science, 1987a), compared with the very explicit progressive child-centred philosophy of the Mathematics Working Group. Perhaps for this reason, but far more likely because of its different significance (the progressive and public educator positions are directly antithetical to that of the industrial trainers in mathematics but not in science), major criticism from Baker did not result. The philosophy became explicit in the Final Report (Department of Education and Science, 1988a). For secondary school pupils 15-20% of the science curriculum was to be devoted to Science in Action (Technological and Social Aspects and The Nature of Science); for all pupils 65% tapering down to 40% was to be devoted to process aspects of science (Exploration & Investigation and Communication); only 35-40% was devoted to traditional Knowledge and Under-standing. These can roughly be seen as corresponding to the public educator aims, the progressive educator aims, and the other three sets of aims, respectively.

Baker's response was to ask for the other components to be combined with those for Knowledge and Understanding, although he allowed that there could be a second profile component (**op. cit.**). This represented a shift towards the aims of the three more reactionary groups (old humanist, technological pragmatist and indus-trial trainer) philosophy was clearly at work), resulting in 50% growing to 70% of the curriculum being devoted to Knowledge and Understanding. The complemen-tary curriculum weighting, 50% tapering down to 30%, is devoted to Exploration

Figure 2
The shift in composite aims of the National Curriculum
in science during its development

AIMS OF THE IDEOLOGICAL GROUPS

```
                Public      Progressive    O.H./T.P.      Industrial
                Educator    Educator       Alliance       Trainer
                  |           |              |              |
Sequence of       |           |              |              |
Documents

     INTERIM-----  XXXXXXXXXXXXXXXXX
     REPORT       XXXXXXXXXXXXXXXXXXXXXX
                 XXXXXXXXXXXXXXXXXXXXXXXXXX
     FINAL------- XXXXXXXXXXXXXXXXXXXXXXXXX
     REPORT       XXXXXXXXXXXXXXXXXXXXXXXXXXX
                   XXXXXXXXXXXXXXXXXXXXXXXXX
     CONSULTATION-            XXXXXXXXXXXXXXXXXXXXXXX
     DOCUMENT                 XXXXXXXXXXXXXXXXXXXXXX
                              XXXXXXXXXXXXXXXXXXXXX
     PARLIAMENTARY-           XXXXXXXXXXXXXXXXXXXXXXX
     ORDERS
```

of Science. This is practical, largely skills-based scientific work, a traditional part of the science curriculum, with the more progressive educator-orientated components on investigation and communication proposed in the final report stripped away (**op. cit.**). Thus the fact that a greater weighting is assigned to practical work in science than in mathematics does not appear to signify any greater impact or retention of the progressive educator aims. It should also be noted that the Science Working Group's bid for 20% of the whole school curriculum failed, and only 12.5% Science is mandatory.

CONCLUSION

The development of the National Curriculum in mathematics, as recounted above, indicates that a conceptual change view of mathematics was never considered by any of the protagonists, since the public educator view was altogether unrepresented. This is not true in the case of the science curriculum, but there the conceptual change view of scientific knowledge was eliminated from the National Curriculum in science. The progressive absolutist view of mathematics was might be said to underlie the strong progressive educator influence on early drafts of the National Curriculum in mathematics (and science), but was also eliminated, or at least reduced to vestigial form in the final, legally enshrined version. Thus the National Curriculum in mathematics embodies an absolutist view of mathematical knowledge, which sees it as a fixed and certain body of objective knowledge cast in a rigid hierarchical form. (The same is largely true of science).

The above comparison also shows that although there was more overt conflict and heavy-handed exercise of power in the development of the National Curriculum in mathematics than in science, the outcomes are broadly similar. No trace of the public educator philosophy survives in either of the two National Curriculum subjects, and little of the progressive educator philosophy. Thus an alliance of the other three groups seems to have largely had their way, each making some concessions to the others. The dominant industrial trainers really wanted a back-to-basics core curriculum for the masses, coupled with a more advanced curriculum for a select minority. They have not wholly succeeded, although the recent decision to restrict national assessments in the primary school to the core subjects is a further move in this direction. When the nature of the assessments to be used is made public, a final crucial indicator of the underlying philosophy of the National Curriculum, and the extent of the industrial trainer domination of it, will be evident. Meanwhile, the industrial trainers have wholly succeeded in imposing the market forces model (coupled with central regulation) on education, including its central curricular feature: the imposition of an assessment driven curriculum, with the curriculum made up of traditional subject areas represented purely as hierarchies of assessment objectives, to be tested nationally and publicly. One outcome of this is likely to be the narrowing down of pedagogical approaches and educational aims in the direction of these tests ('teaching for the test'), a triumph for the educational aims of the industrial trainers. Another outcome is likely to be a hierarchical ordering of schools by test results, popularity, resourcing, and ultimately by the wealth and social class of their clientele. Through such processes the outcome of market forces competition in education will be a 'pecking order' of schools, reflecting the industrial trainers aims and model of society, and fracturing the remnants of the comprehensive ideal and what exists of equality of opportunity in education. This would represent a triumph for the social aims of the industrial trainers.[3]

A final issue that remains outstanding concerns the apparently contradictory nature of the ideologies of the groups which formed an alliance in the National Curriculum in mathematics. The old humanists have a strongly held purist view of mathematics, whilst the technological pragmatists and industrial trainers are concerned with the utility and applications of mathematics. How can such contradictory views admit alliances and cooperation? A plausible answer is that there is more in common between these groups that at first meets the eye. One immediate fact is that the contradictory aims of these three groups do not all apply to the same sector of the population. The industrial trainers require basic skill and 'social training in obedience' for the masses, represented by the average and below average mathematical attainers in schools. These pupils, because of the well known association between attainment and class in British schools, do largely represent the 'masses' and working classes. The old humanist aims of education in pure, theoretical mathematics apply to high attainers, who continue on to university/ higher education and make up the future professional and middle classes. Thus the constituencies of concern of the industrial trainers and old humanists are largely mutually exclusive and complementary. The Technological pragmatists aims are differentiated according these two main groups, and in each case attempt to soften the harder aims of the other two groups: aiming to upgrade the competencies of workers and to apply the theoretical knowledge of some professionals through their occupations. Thus the conflict is actually far less than it appears at first sight.

It can also be remarked that the old humanists as a group have actually had very little direct impact on the National Curriculum in mathematics, except to 'beef up' some of the content of the higher levels of the attainment targets. What is perhaps far more important is that each of these three groups shares an absolutist philosophy of mathematics. Mathematics is perceived by each of the groups to be objective, incorrigible and neutral. This philosophy also leads to a hierarchical view of mathematical knowledge, as is manifested in the structure of the National Curriculum in mathematics (and indeed in all subjects). This characteristic can be linked with hierarchical views of human nature and mathematical ability, and indeed with hierarchical views of society (Ernest, 1991). This is evident in the above account of the aims of the groups applied differentially to different social sectors. Such a socially based view of the aims of these three groups suggests that they are united in yet another aim. This is the preservation of the social *status quo* and hence their own social privileges (Ernest, 1990, 1991). For each of the three groups is privileged in the power and resources it commands. Thus ultimately, each of the social groups may be seen as sharing similar social interests, which their alliance allows them to preserve. Although speculative, such a conclusion does have a ring of plausibility to it, being consistent with a range of other well-known analyses of education in terms of social reproduction.

The conclusion that follows from the analysis offered in this paper is that the National Curriculum is the outcome of vested interests and political ideologies, and the aims it expresses cannot be said to spring from a properly articulated educational philosophy, let alone the love of children, the love of mathematics or a deep wish to build a better society.

POSTSCRIPT 1992

Since this paper was written in 1989, developments reveal a yet further shift of power. To extend the analysis offered above, this indicates that the Industrial Trainers have grown in confidence, and are dominating any expedient alliances they formed with Technological Pragmatists and Old Humanists. Thus members of New Right think tanks are now installed in positions of power, rather than sniping from the sidelines. Lord Griffiths is chair of the School Examinations and Assessment Council (SEAC); Dr. J. Marks chairs its mathematics committee, and has just appointed Professors S. Prais and G. Howson to that committee. Others, such as Caroline Cox and Professor R. Scruton have been given other official roles. Outside education, Lord Shawcross, the most prominent far-right critic of the media for its so-called 'left wing bias' has been appointed as chair of the official media watchdog.

Major revisions of education in line with the Industrial Trainer perspective are in place, currently underway, or planned. The Task Group on Assessment and Testing proposals for a wide range of testing modes including practical testing is being scrapped, and paper and pencil tests substituted at 7, 11 and 14. Coursework in GCSE examinations at 16+ is being severely cut back. The National Curriculum in Mathematics and Science have been simplified, reducing the progressive educator elements in both (and eliminating the socio-historical vestiges of the Public Educator perspective from Science). Further Education colleges and polytechnics have been privatized, stripping Local Education Authorities of further powers. However a new bill will give central government the power to withhold funds from universities if they consider that they have good reason. These are the first legal powers ever to curb academic free speech and inquiry.

Any opposing rational argument is being squashed. The chairs of National Curriculum Council and SEAC being replaced by yes-men or ideologues. The HMI are being severely cut and privatised, they are being penalized for their steadfast (if somewhat conservative) integrity. Teacher educators are under severe threat as 'progressive' enemies of the New Right and 'ideological poisoners' of new teachers. Teacher education is to be largely school based, with an emphasis on basic skills, authoritarian discipline, and traditional teaching, according to the appropriate minister: Kenneth Clark. The senior civil servants concerned with teacher education, arguing from a Technological Pragmatist perspective that this latter reform will prove expensive and may be counterproductive, have been moved sideways. On such a strong ideological issue, financial considerations are apparently given second place by the Industrial Trainers.

Although Margaret Thatcher has been out of power for over a year, the radical reforming zeal she represented is gaining in momentum. Kenneth Clark, fresh from partially privatizing the National Health Service, is applying his Industrial Trainer view to education with great gusto. There are some contrary developments, such as widening access to Higher Education, which sits better with progressive ideologies than with that of the Industrial Trainers. But this latter group do have anti-elitist tendencies manifested in their onslaughts on professional autonomy (and attacks on Royalty) which might accommodate this seemingly progressive development.

In retrospect, the only criticism I might direct at the above analysis is to ask why the checking of the Industrial Trainers was not better accounted for? What forces held them back? Certainly they are now pursuing their educational and social policies with a new boldness and vigour, unfettered by compromises with other groups. Perhaps the Establishment had a built-in conservatism which resisted radical change while the climate of liberal argument and rationality persisted. Perhaps there was a traditional middle class veil of politeness and compromise masking and softening the realities of the exercise of power, which has now been removed. Certainly the velvet glove has now come off the iron fist. If this analysis is right, then it may be a long time before a balance is restored in the governance of education in Britain, let alone a move to a Public Educator position.

NOTES

1. This provides an account of the historical development of each of the five groups, and gives an explicit treatment of the ideology of each of the groups in term of two sets of components. The primary elements are epistemology, philosophy of mathematics, set of values, theories of society and of the child, and aims for education. The secondary elements, concerning mathematics teaching, are theories of: school mathematics, aims, teaching, resources, learning, ability, and social diversity.

2. Additions of higher mathematical content were made to the upper parts of Attainment Targets, for example, to AT 7 at Level 7, AT 10 at Levels 9 & 10, AT 11 at Level 6, AT 12 at Levels 9 & 10, AT 13 at Level 10 and AT 14 at Level 10. The consultation process reported in National Curriculum Council (1988) attracted old humanist criticism, such as that in the response of the Mathematical Association (1988), which implicitly attacked the progressive philosophy ("AT 15 is of a different nature...[personal qualities]...are not specific to mathematics"), and promoted the mathematics-centred view "We wish ... proof ...[to] be included... Is there a case for the inclusion of matrices...?" **(op. cit.,** p. 5).

3. These pessimistic conclusions are not inevitable. A new government with different social and educational goals can reverse them. Teachers can also subvert the industrial trainer aims by empowering learners with a 'problem posing pedagogy' (Ernest, 1991).

4. Professor Howson is no member of the New Right, but he may be deemed admissible by them because of his critique of the National Curriculum published by the Centre for Policy Studies, which called *contra* TGAT for paper and pencil tests of core subject skills.

REFERENCES

Aleksander, I   (1988)   Putting down the professionals,
New Scientist, 25 February 1988, p.65.

Abraham, J.
and Bibby, N.   (1988)   Mathematics and Society: Ethnomathematics and the
Public Educator Curriculum, *For the Learning of
Mathematics*, Volume 8, No. 2, 2-11.

Bash, L.
and Coulby. D.   (1989)   *The Education Reform Act,*
London: Cassell.

Beck, J.   (1981)   Education, Industry and the Needs of Industry,
Education for Teaching, Volume 11, No. 2, pp.87-106.

Bell, A.W.,
Costello, J. and
Kuchemann, D.   (1983)   *A Review of Research in Mathematical Education (Part A),*
Windsor: NFER-Nelson.

Benecerraf, P. and
Putnam, H.
(Eds)   (1964)   *Philosophy of Mathematics,*
Englewood Cliffs: Prentice-Hall.

Blenkin, G. and
Kelly, A.V.   (1981)   *The Primary Curriculum,*
London: Harper & Row.

Bloor, D.   (1976)   *Knowledge and Social Imagery,*
London: Routledge & Kegan Paul.

Brown, M.   (1988)   Teachers as Workers and Teachers as Learners,
                              presented at *Sixth International Congress of
                              Mathematics Education,*
                              Budapest, July 27-August 4, 1988.

Campaign for
Real Education (1987)   *Campaign for Real Education Credo,*
                              York: Campaign for Real Education.

Chisholm, R.   (1966)   *Theory of Knowledge,*
                              Englewood Cliffs: Prentice Hall.

Chitty, C.   (1987)   The Commodification of Education
                              Forum, 24(3) 66-68.

Cockcroft, W.H.
(Chair)   (1982) *Mathematics Counts,*
                              London: Her Majesty's Stationery Office.

Confrey, J.   (1981)   Conceptual Change Analysis: Implications for
                              Mathematics and the Curriculum,
                              Curriculum Inquiry,
                              Volume 11, Number 5, pp. 243-257.

Cusin, B.   (1972)   *Ideology,*
                              Milton Keynes: Open University Press.

Cox, C. and
Marks, J.   (1982)   *The Right to learn,*
                              London: Centre for Policy Studies

Cox, C.B. and
Boyson, R. (Eds)   *Black Paper 1975: The Fight for Education,*
                              London: Dent and Sons.

Crane, D.   (1972)   *Invisible Colleges,*
                              Chicago: University of Chicago Press.

Davis and
Hersh   (1983)   *The Mathematical Experience,*
                              Harmondsworth: Penguin Books

Dearden, R.F.   (1968)   *The Philosophy of Primary Education,*
                              London: Routledge and Kegan Paul.

DES   (1987)   *Interim Report of the Mathematics Working Group,*
                              London: Department of Education and Science.

DES	(1987a)	*Interim Report of the Science Working Group,* London: Department of Education and Science.
DES	(1987b)	*The National Curriculum: A Consultation Document,* London: Department of Education and Science.
DES	(1988)	*Final Report of the Mathematics Working Group,* London: Department of Education and Science.
DES	(1988a)	*Final Report of the Science Working Group,* London: Department of Education and Science.
DES	(1988b)	National Curriculum: Task Group on Assessment and Testing: *A Report*, London: Department of Education and Science.
DES	(1989)	*Mathematics in the National Curriculum,* London: Department of Education and Science.
DES	(1989a)	*Science in the National Curriculum,* London: Department of Education and Science.
DES	(1989b)	*National Curriculum: From Policy to Practice,* London: Department of Education and Science.
Ernest, P.	(1990)	The Aims of Mathematics Education as Expressions of Ideology, in Noss et al. (1990), pp.87-94.
Ernest, P.	(1991)	*The Philosophy of Mathematics Education,* London: Falmer Press.
Golby, M.	(1982)	Computers in the Primary Curriculum, in Garland, R. (Ed) (1982) *Microcomputers and Children in the Primary School,* Lewes: Falmer.
Gordon, P.	(1989)	The New Educational Right, *Multicultural Teaching,* Volume 8, Number 1, pp.13-15.
Gow, D.	(1988)	Adding up to numeracy, *The Guardian,* 23 February
HMI	(1977)	*Curriculum 11-16,* London: Her Majesty's Stationery Office.

HMI   (1985)   *Mathematics from 5 to 16*,
                            London: Her Majesty's Stationery Office.

Heyting, A.   (1956)   *Intuitionism: An Introduction*,
                            Amsterdam: North Holland.

Himmelfarb, G. (1987)   *Victorian Values*,
                            London: Centre for Policy Studies.

Keitel, C. with
Damerow, P., Bishop, A.
and Gerdus, P. (Eds)
                (1989)   *Mathematics, Education and Society*,
                            Paris: UNESCO.

Lakatos, I.   (1976)   *Proofs and Refutations*,
                            Cambridge: Cambridge University Press.

Lakatos, I.   (1978)   *Mathematics, Science and Epistemology*
                            (Volume 2), Cambridge: Cambridge University Press.

Lawlor, S.   (1988)   *Correct Core*,
                            London: Centre for Policy Studies.

Lawton, D.   (1984)   *The Tightening Grip*,
                            London: Institute of Education.

Lawton, D.   (1988)   *Ideologies of Education*, in Lawton and Chitty
                            pp10-20.

Lawton, D. and
Chitty, C. (Eds) (1988)   *The National Curriculum*,
                            London: Institute of Education.

Letwin, O.   (1988)   *Aims of Schooling*,
                            London: Centre for Policy Studies.

Maw, J.   (1988)   National Curriculum Policy: coherence and progres-
                            sion? in Lawton and Chitty (1988), pp.49-64.

Mayhew, J.   (1987)   *Statements about mathematics and mathematics educa-
                            tion, Some possible consequences for the National Math-
                            ematics Curriculum*, unpublished paper, dated 7 Sept.

Meighan, R.   (1986)   *A Sociology of Educating*,
                            Eastbourne: Holt, Rinehart and Winston.

National Curriculum
Council   (1988)   *Consultation Report: Mathematics,*
York: National Curriculum Council.

            (1988a)   *Consultation Report: Science,*
York: National Curriculum Council.

            (1989)   *Mathematics: Non-Statutory Guidance,*
York: National Curriculum Council.

            (1989a)   *Science: Non-Statutory Guidance,*
York: National Curriculum Council.

Noss, R.   (1990)   The National Curriculum and Mathematics: a case of
divide and rule? paper presented at *RSPME conference,*
Institute of Education, June 1989.

Noss, R., Brown, A., Drake, P., Dowling, P., Harris, M.,
Hoyles, C. and Mellin-Olsen, S. (Eds)
            (1990)   *Political Dimensions of Mathematics Education: Action
and Critique,*
London: Institute of Education.

Perry, W.G.   (1970)   *Forms of Intellectual and Ethical Development in the
College Years: A Scheme,*
New York: Holt, Rinehart and Winston.

Prais, S.J.   (1987)   *National Curriculum Mathematics Working Group: In-
terim Report, Note of Dissent,* 30 November 1987,
London: National Institute of Economic & Social
Research.

Prais, S.J.   (1987a)   *Further response to National Curriculum Mathematics
Working Group,* 10 December 1987, unpublished
paper.

Pring, R.A.   (1988)   personal communication.

Raban, J.   (1988)   God, Man and Mrs. Thatcher,
The Observer, 28 May 1989, p.33.

Ramsden   (1986)   'Ideologies of the Child', *unpublished course notes,
University of Exeter School of Education.*

Scott-Hodgetts, R.(1988)   The National Curriculum: Implications for the Sociology of Mathematics Classrooms, presented at *RSPME Conference*, Polytechnic of the South Bank, London, December 1988.

Thatcher, M.   (1977)   *Address to the Conservative Party Conference,*   October 1987,
London: Conservative Central Office.

Tymoczko, T. ed.(1986)   *New Directions in the Philosophy of Mathematics,* Boston: Birkhauser.

Warnock, M.   (1989)   *Universities: Knowing our Minds,*
London: Chatto and Windus.

Williams, R.   (1961)   *The Long Revolution,*
Harmondsworth: Penguin Books.

Wittgenstein, L. (1978)   *Remarks on the Foundations of Mathematics,*
Cambridge, Massachusets: The MIT Press.

Young, H.   (1989)   *One of Us,*
London: Macmillan.

Young, M. F. D. (Ed)
    (1971)   *Knowledge and Control,*
London: Collier-Macmillan.

Woozley, A. D. (1949)   *Theory of Knowledge,*
London: Hutchinson.

THE NATIONAL CURRICULUM AND MATHEMATICS: POLITICAL PERSPECTIVES AND IMPLICATIONS

Richard Noss

I outline here the basic arguments proposed in a paper originally presented to the RSPME group. It has since been published under its original title: The National Curriculum and Mathematics: a Case of Divide and Rule? in : Dowling P. and Noss R. (Eds), 1990. Mathematics versus the National Curriculum. Basingstoke, UK: Falmer Press.

INTRODUCTION

I shall argue that although the underlying purpose of the National Curriculum and the testing procedures which accompany it are profoundly and intentionally anti-educational, they are permeated by contradictions which derive from their origins; and that by exploiting these contradictions it may be possible for teachers to use some of the proposals for educational purposes - effectively to subvert the implicit and explicit intentions of those responsible for their introduction.

It is helpful to examine the working party reports which preceded the bare attainment targets of the final curriculum. The basic assertions contained in these are:

(i)   that a centrally imposed system of curriculum and assessment will bring about the increase in 'standards' which the authors desire; and

(ii)   that the TGAT (the Task Group on Assessment and Testing) proposals are valid and workable as a means of delivering it.

It is my belief that neither of these assumptions is valid. However, my intention here is to take up a more general question altogether; to ask whether the expressed intention of 'raising standards' really lies at the hear of the proposals, and to discuss what actually constitutes the rationale of the National Curriculum and its attendant testing procedures.

CONTRADICTIONS

In the working party reports, contradictory positions abound. Assertions extolling mathematics as a creative and aesthetic subject coexist with statements such as: 'We have taken it as axiomatic that the mathematics which pupils learn at school should support the mathematics which they actually need to use in later life, particularly at work'. (Maths 5-16; para 3.150. In fact, the view of mathematics which permeates the National Curriculum is based on the metaphor of a tool: having selected the appropriate mathematics, pupils should 'apply it sensibly and efficiently, try alternative strategies if needed, check on progress at appropriate stages and analyse the final results to ensure that the initial requirements have been met'. Here is the language of the workplace: mathematics is a means for working people to solve the problems of their employers.

Alongside this view, there exists a hierarchical view of mathematics itself, and its associated pedagogy. Thus justifying why it started with attainment targets rather than with defining programmes of study, the working party states: 'We suggest that our experience very largely reflects the nature of mathematics. Attainment targets in mathematics have to be very tightly defined to avoid ambiguity, and the degree of precision required gives a very clear indication of the content, skills and processes associated with the targets' (Maths 5-16; para. 4.15). The view that mathematical concepts can indeed be formed into a rigid hierarchy is open to question on both epistemological and psychological grounds.

Imposing a coherent perspective onto an essentially incoherent set of documents is potentially misleading, and does not take us to the heart of the explanation we seek. So, for example, we have witnessed the introduction of long multiplication at precisely the point in history where the application of the algorithm has become historically redundant thanks to the calculator. Thus an appeal to utilitarianism as the actual rationale is too simplistic: we need to look elsewhere to explain its reintroduction into pupils' mathematical experience.

FORM IS THE PRIORITY
My thesis is that the content of the National Curriculum provides nothing more than a thin veneer for its essential form: that the National Curriculum is about testing and grading *per se*, and that what is tested is of less than secondary importance. This relative lack of governmental concern for the content of its National Curriculum is a corollary of a more general phenomenon: namely that schooling performs primarily a social rather than a training function for its pupils. I argue that the changes in educational orthodoxy have not been arbitrary; they are not solely the result of the whim of a reactionary government. To understand the process it is necessary to look beneath the surface of both liberal and reactionary rhetoric, to the underlying economic and social changes which have occurred and the impact of these on the priorities of the educational system.

Harry Braverman, in his book Labour and Monopoly Capital puts the general case eloquently: '...it is the FORM of the educational encounter - the social relations of education - that accounts for both its capacity to reproduce capitalist relations of production and its inability to promote domination and healthy personal development.' This general perspective finds its mathematical corollary as follows: '[The programmes of study] ...seek to draw out each child's full potential through the development of sound work habits, self-discipline and industry together with good personal qualities' Maths 5-16, para. 8.4.

It is worth noting that the advent of new technology - often cited as the central rationale for a utilitarian perspective, and as the reason for the tight specification of mathematical attainment - has in fact produced precisely the reverse effect. The central impetus of twentieth century capitalism has been the 'de-skilling' of the labour process. Rapid changes in technology have created an impressive array of new jobs to act essentially as minders for machines - and the system itself has actually required and used the machines) to gain ever more control of the labour process at the expense of craft knowledge and employee control. The individual in Western capitalism needs to know less and less mathematics to function effectively in the work place.

The perspective outlined here suggests that the naive optimism of many within the mathematics education community is misplaced. It may be true that the acceptance of the TGAT structures and the incorporation of crumbs of 'progressive' educational thinking into the final orders, represents a victory over the 'theorists' of the Centre for Policy Studies who are urging much more draconian measures. But in fact, arguing about the niceties of the content within individual statements of attainment or the supposed equivalences across levels, diverts attention away from the specific form of the curriculum: and it is this which will primarily distort the learning and teaching of mathematics, and of the educational system itself.

CONCLUSION

The targets and programmes of study represent an educational vacuum. My claim in this paper is broadly that even within the confines of the National Curriculum and centralised assessment, there is room for teachers to teach and for children to learn; that content - precisely because it is such a low priority for those involved in the imposition of the National Curriculum - can be raised to a high priority by teachers. In arguing this case, I am aware that I run counter to some mathematical orthodoxies which seem in the recent past to have implicitly or explicitly advocated a policy of what amounts to disguising mathematical content in games, puzzles and investigations; rightly emphasising the importance of mathematical processes but wrongly, in my view, neglecting the need to help pupils to come into explicit contact with mathematical ideas.

Part 2

THE SOCIAL CONTEXT OF MATHEMATICS EDUCATION

INTRODUCTION

Until recent years the main concern of mathematics educators was what to teach, and when and how to teach it. Whilst all education is socially situated, in its widest sense, mathematics education, dealing as it does with apparently timeless and abstract objects, procedures and knowledge, was as little concerned with the sociology of the classroom as possible. There has always been a recognition that conditions in the classroom need to be conducive to learning, but these were background issues. Just as 'true' mathematics is seen to need no justification, whereas there can be a sociology of errors in the development of mathematics, so too there can be interference in the process of learning, and this is a legitimate concern of the mathematics teacher, but eliminating the interferences will inevitably enable pure learning to take place. However, new analyses, particularly in the last decade, have changed the emphasis to a recognition that mathematics itself is as socially embedded as any other form of knowledge, and since so too are the cognitive and the affective, in relation to learning and teaching mathematics, we have a great deal of work to do in this area.

The papers in this section deal with three particular aspects of the social context of the mathematics classroom: language; teachers' expectations of cognitive development and student-teachers' beliefs.

David Pimm considers the ways in which verbal interactions established by the teacher, in the classroom, permit or exclude communication. Teachers inadvertently employ overt and covert moves to effect these frameworks. Unless someone holds up a mirror to our actions, a mirror that offers possible interpretations of what we see, we perpetuate those moves that are often, in fact, replications of our own teachers' behaviours. Pimm attempts to perform an interpretive mirror function in his paper "Classroom Language and the Teaching of Mathematics".

Richard Winter invites us to look again at the effects of our understanding of cognitive development in children. He draws on interviews with young children to question the Piagetian notion that understanding of infinity comes at the formal operational stage of development, and suggests that the limitations we impose on children by assumptions of "ages and stages" may well lead to loss of confidence and fear of mathematics, in his paper "Mathophobia, Pythagoras and Roller-Skating".

Barry Cooper examines the influence of student-teachers' beliefs on their actions as teachers, in his paper "PGCE students and investigational approaches in secondary maths". The move towards investigations in schools had been gradual, until it became clear that there was to be the introduction of coursework into GCSE mathematics. At many universities and colleges, however, investigational work became part of PGCE courses quite early on, for a variety of reasons which Cooper examines. He then follows four student teachers in depth, through interviews and observations, to attempt to identify some of the connections between beliefs and practice, in the context of their attitudes to investigations.

CLASSROOM LANGUAGE AND THE TEACHING OF MATHEMATICS

David Pimm

FOREWORD

I presented a paper to the group in 1986, entitled 'Aspects of overt and covert classroom communication in mathematics'. This piece subsequently became part of Chapter 3 in my book *Speaking Mathematically* (Pimm, 1987) on mathematics and language. Rather than merely reprint that article directly, I have chosen to describe briefly the orientation of that piece and then to develop some of the ideas with further examples and analysis.

In that chapter, I introduced the notion of a teaching 'gambit'. One feature of a gambit is that there is some sacrifice involved in order to move toward a hoped for long-term advantage. I exemplified the notion of gambits by three situations: having pupils talk in pairs, not answering a question directly and use of silence. I then went on to look at common ways of asking, answering and deflecting questions in mathematics classrooms with the tacit focus on what some of the sacrifices might be as well as the hoped-for advantages. The piece ended with an exploration of the potential interpretations of the widespread use of the word 'we' by teachers of mathematics.

The focus throughout the article was on specific aspects of classroom interactions between teachers and pupils (particularly spoken ones) and their significance, highlighting in certain circumstances the interesting or incongruous usages that pertain in classrooms, particularly when mathematics is under discussion.

In the remainder of this article, I want to pursue similar themes, specifically that of teacher gambit and intention as embodied in particular discursive strategies in the contexts of questions and answers, and of whole-class report-backs, an increasingly common feature not only of 'investigation lessons', but many mathematics lessons.

There are many different aspects of and relationships between mathematics and language that can be highlighted as part of the mathematics education enterprise. My intention in this article is to start to explore some of the particularities of teacher-pupil spoken interactions, but I feel that it is of crucial importance to find ways of talking about the varied components of mathematical activity itself. One means for achieving this would be to focus attention on particular features of doing mathematics which might then afford teachers greater insight into what is happening in and between their pupils when in the mathematics classroom. This article, and indeed this volume, focuses particularly on linguistic aspects of doing and teaching mathematics, in particular the social context and community which classrooms exemplify.

REPORTING BACK TO THE WHOLE CLASS ON INVESTIGATIONS

Eric Love (1988) has produced a thoughtful account of the development of the last thirty years and analysed the variety of ways people came up with to try to describe mathematical activity beyond a purely content description, and in particular on the birth of the noun 'an investigation'. He comments:

> In the early writings on mathematical activity, there is no mention of 'investigation' in the sense of 'doing an investigation'. It is interesting to see this construction develop from pupils 'investigating such-and-such', or 'carrying out an investigation into such-and-such', no doubt originally as a shorthand, but soon taking on a life of its own. The path to formalisation had begun.
>
> In the last decade 'investigations' have become institutionalised - as part of formal requirements for assessment of courses ... They also appear in the official recommendations of Cockcroft and HMI.
>
> Such a development is a typical one in education - the often commented upon way in which originally liberating ways of working become formalised and codified, losing their purpose as they become adapted for different ends or by those who have no personal commitment to the underlying intentions. (pp. 249-51)

One example of such 'institutionalisation' has been that post-Cockcroft 243 (HMSO, 1982, para. 243), there has been the emergence of the phrases the 'investigation lesson' or the 'discussion lesson' in the same way as the terms the 'algebra lesson' or the 'decimals lesson' would formerly have been used to describe the main focus. Such a usage also gives another example of this process of action turning into noun, whereby everything ends up as the content to be taught in a lesson. These expressions seem to me to mistake the means for the end - the end is not just discussion *per se*. I have written elsewhere (Pimm, 1986, 1987) on the naïve assumption that pupil discussion is wholly and everywhere a good thing, rather than being a particular tool with strengths and weaknesses available for use in various situations.

The 'investigation lesson' has its common structure of whole-group posing of the task, pupils then working together in small groups and then reporting back to the whole class. This format has become an example of the new orthodoxy, and, unexamined, can be as harmful to the health of teacher-pupil communication as the old pattern of the teacher-led, blackboard-focused lesson style. Nonetheless, the activity of reporting back throws up some interesting linguistic questions, in relation to teacher and pupil language and styles of speech in the context of the mathematics classroom.

Because of the more formal nature of the language situation (particularly if rehearsal is encouraged), reporting back can lead to more formal, 'public' language being used and greater structured reflection on the task. Thus, the demands of the situation alter the requirements of the language to be used. With regard to preparation for coursework write-ups, for example, a prior stage of oral reporting back can help with selection of material. and the emphasis placed on various parts of it

One key question is for whose benefit is the reporting back, one which may be particularly pertinent if you have been present at sessions that were apparently very painful for some of the participants. Requesting pupils to report back can also be intrusive and perhaps unhelpful if some groups are not at the stage for summarising what they are have done and are still engaged with the problem. Before reading on, you may care to think for yourself about the question of who benefits (if anyone) and why.

There seem to me to be at least three possible answers and corresponding sets of potential justifications.

(a)   **The pupil(s) doing the reporting**: for them, plausible justifications include development of a range of communication skills, their use of language and development of social confidence. A further important possibility is that of developing skill at reflection on a mathematical experience, and distilling it into a form whereby they as well as others may learn from it.

(b)   **The other pupils listening**: for them, potential benefits include hearing alternative approaches and that perhaps others besides themselves had difficulties, the chance to ask genuine questions of other pupils and to engage in trying to understand what others have done on a task that they also have worked at.

(c)   **The teacher**: potential benefits here include a range of opportunities to make contextually-based meta-remarks about methods, results and processes (perhaps indicating the task is open and that a number of ways of proceeding are possible or different emphases can be placed), as well as to value pupils' work and broaden pupils' experience (possibly by sharing an original idea with the rest of the class). "Did anyone else ..."

Reporting back can also provide the teacher with access to what the pupils think the task was about, as well as less tangible but nonetheless interesting information such as what they think is important (in terms of what they select to talk about) and their level and detail of oral expression of mathematical ideas. But it can also be *assumed* that reporting back *has* to happen, perhaps meeting the need (or expectation) for some sort of whole-group ending, 'pulling the session together'. If that is the case, then many of the above benefits may not necessarily accrue, because attention is not being paid to their elicitation.

Here are five further questions that, for me, encapsulate some of the key issues involving mathematics education and language arising from the task of reporting back on mathematical investigations.

(1)   *What sort of language (style, structure, organisation, register, and so on) does the reporter use?*

How can the teacher help resolve the tension between the fact that increased preparation time can improve performance, but also can serve to emphasise the public speaking aspect of the event, with its attendant social pressures on performance? However, spontaneous requests without much time for reflection or organisation, though offering a potentially less formal context for speaking, can merely result in recollections of the last thing they were doing.

(2)   *How can the tension between wanting the pupil(s) to say themselves what they have done, while wanting to use what they say to make general remarks about how to undertake investigative work be contended with?*

This can be particularly strongly felt in the case where the teacher has seen something that they feel is an instance of a higher-order process that they value (be it specialising systematically, developing notation, coping with getting stuck, or whatever) while circulating around the small groups. When invited to tell the class about this incident, it may well have not been a salient one for the pupil (unless the teacher made a big point of it at the time), so they probably have little idea either of what to emphasise or why this particular incident is being focused on. If the teacher has made the point to them, why are they being asked to repeat it?

There seems to be a reluctance for teachers to say for them what one of the pupils has done, instead of prompting with something like: "Why don't you say it, because you did it?" A truthful response from the pupil might be: "Because it is *your* anecdote. I don't know what significance it has *for you*". They do not have the teacher's reasons for highlighting particular parts of what they have done. Far better possibly to say something like: "I want the rest of the class to know about something related to what you did. At the moment, I am the only one who knows what that is. What I saw you doing was ..., and it struck me because ...".

(3)   *How can pupils develop the linguistic skills of reflection and selection of what to report? How can they work on acquiring a sense of audience?*

Obviously, this is contingent on the perceived audience and purpose behind carrying out the reporting back. Who knows what might be worth telling about? The teacher has no control over what comes out. He or she can only work on it after it has emerged from the pupil, though it is likely that what they then do with the response will influence later reporters.

One important skill is the ability of pupils to disembed their discourse from the knowledge of the group who saw the work developed, in order that someone who was not there can follow what is being described. The tendency is for the reporter to assume that everyone will know what they are talking about.

(4)   *To whom is the reporter talking?*

Pupils seem often to address themselves to the teacher, the person who, besides their own group, probably knows *most* about what they have done - and the pupils know this. How might the teacher deflect the reporter's attention to the rest of the class?

In Barbara Jaworski's account (1985) of a poster lesson she has observed, she comments:

> Most pupils, on taking the hot seat started off address-
> ing their comments to Irene [the teacher], and stop-
> ping now and then in what they were saying to allow
> her to comment, or to solicit her comments. In some
> cases she replied to them or prompted them directly,
> which then encouraged a two-way exchange with the
> rest of the class as audience. In order to get the rest of
> the class to participate she then had to overtly invite
> comments from them. What in fact started to encour-
> age more general discussion was Irene's deflecting of
> the invitation to comment to others in the group.

If the teacher reinterprets what the pupil says for the rest of the class (possibly by playing a role which in television interviews is known as 'audience's friend'), what effects might this have on the reporter? By playing 'audience's friend', the teacher can also ease the strain on the reporter, by taking the focus off them, possibly by reinterpreting or expanding for the audience and then moving into more general questioning of the class.

However, it also may result in the teacher being looked to as broadcaster and interpreter of the person reporting back and thus acting as an intermediary between reporter and audience which may get in the way. After all, if reporting back were an effective technique in and of itself, there would be little need for the teacher to intervene - the reporting back would do its own work.

If the teacher has adopted this role, then she may be asking questions on behalf of the audience or more directly explaining the pupil's words to the others. Whether the teacher says "Tell us ..." or "Tell them ..." may be an important difference in terms of cueing the reporter as to whom they should be facing and speaking.

Where is the teacher standing and where is the control? If there is a silence, whose responsibility is it to fill it? Where is the audience's attention and who are they asking questions of? If it is predominantly a conversation between the reporter and the teacher, what is the intended role for the pupils who are being invited to listen?

What would you (as their teacher) like to be happening in their heads and what could you do who help bring it about? What techniques do you employ for deflecting or involving the audience? What are the pupils supposed to be doing while the report-back is being given? Are they too concerned with the fact that their turn is coming up (and are perhaps rehearsing what they are going to say) to attend to what the current reporter is saying? How can they be encouraged to be *active* listeners?

(5)   *What justifications does the teacher explicitly offer the class for reporting back on their work to the rest of the class?*

What covert justifications does the teacher have? What are various pupil views about why they have been asked to engage in this activity? Do they mirror what the teacher has said?

There can be some difficulty in conveying to pupils what it is they are being asked to do: if too vague, the pupils do not know why they are being asked to carry it out and can flounder; if too precise, pupils will tend to do exactly that, thereby constraining what might happen. Over-specification also serves to retain the teacher's control and initiative rather than handing this over (at least in part) to the pupils.

This is a familiar tension which lies at the heart of teaching. One general formulation (referred to there as 'the didactic tension') runs as follows (Mason, 1988, p. 33):

> The *more* explicit I am about the behaviour I wish my pupils to display, the more likely it is that they will display that behaviour without recourse to the understanding which the behaviour is meant to indicate; that is, the more they will take the *form* for the substance.

The less explicit I am about my aims and expectations about the behaviour I wish my pupils to display, the less likely they are to notice what is (or might be) going on, the less likely they are to see the point, to encounter what was intended, or to realise what it was all about.

TEACHING AS META-COMMENTING

What time is it?

It is half-past two.

Thank you.
or
Well done.

In her novel *Up the Down Staircase*, Bel Kaufman tells poignantly of a male English teacher receiving a love letter from a female pupil and the way he copes with this is by correcting the spelling and grammar and returning it to her with a grade. Such an extreme instance of attending to form rather than content merely serves to highlight one important possibility for unusual discourse in classrooms. In an early paper entitled 'Organizing classroom talk', educational linguist Michael Stubbs (1975) offers the notion that one of the characterising aspects of teaching discourse as a speech event is that it is constantly organised by meta-comments, namely that the utterances made by pupils are seen as appropriate items for comment themselves and, in addition, that many of the meta-remarks are evaluative. He comments:

> The phenomenon that I have discussed here under the label of meta-communication, has also been pointed out by Garfinkel and Sachs (1970). They talk of "formulating" a conversation as a feature of that conversation.
>
>> "A member may treat some part of the conversation as an occasion to describe that conversation, to explain it, or characterise it, or explicate, or translate, or summarise, or furnish the gist of it, or take note of its accordance with rules, or remark on its departure from rules. That is to say, a member may use some part of the conversation as an occasion to *formulate* the conversation."

I have given examples of these different kinds of "formulating" in teacher-talk. However, Garfinkel and Sacks go on to point out that to explicitly describe what one is about in a conversation, during that conversation, is generally regarded as boring, incongruous, inappropriate, pedantic, devious, etc. But in teacher-talk, "formulating" is appropriate; features of speech do provide occasions for stories worth the telling. I have shown that teachers do regard as matters for competent remarks such matters as: the fact that somebody is speaking, the fact that another can hear, and whether another can understand. (pp. 23-4)

In Stubbs' paper itself, the examples he uses to exemplify his various types of speech acts undertaken by teachers are all from a lesson involving an English language teacher working with a group of teenage pupils who are non-native speakers. Nonetheless, despite the fact that in such a class matters of language and the form of utterances being obvious foci for attention, I am taken with his ostensible characterisation (in part) of teaching in terms of making meta-comments.

In a group discussion of teaching strategies, Elizabeth Richardson (1967) offered the situation of coming into the classroom to find an obscene drawing written on the board. In addition to the strategies of ignoring the content of the message and rubbing it off, and the confrontational approach of demanding "Who wrote that", is the possibility of commenting on the utterance, by drawing attention to the fact that someone has written on the board and asking "I wonder why they did that". In this way the message on the board becomes an object of study and speculation for the class rather than a direct message to the teacher by an unknown communicant. Richardson writes: "...try to get the class to talk about the feelings or preoccupations or anxieties that give rise to such a demonstration of hostility against an adult." (p. 100)

I am interested in the general question "What is teaching?" I find it interesting and provocative when Stubbs (1983) claims: "In a sense, in our culture, teaching is talking". I want to take this metaphoric suggestion seriously and ask what sort of talking constitutes teaching. The option of attending to form over content as a way of directing attention is the one aspect I wish to focus on here. One link with the previous section is that examples from the teacher acting as 'audience's friend' require precisely such language apparently in the conversation about the conversation itself.

Excerpt 1:

David Cain is working with a group of twelve-year-olds and the task he has set them is to go from the net of a cube, with which they are familiar to one for a solid (which he terms a 'slanty-cube') where each face is a rhombus.

The lesson extract opens with the word 'rhombus' written on the board and the net of a cube drawn. He says to the whole class:

> Therefore, all you've got to do is this - very simple. There's the net of a cube - you've got to make these into rhombuses and it should stick together to make a ... slanty-cube. Yes, and instead of a cube, you'll have what is called a rhomboid - right?

Later in the lesson, he is talking with one particular girl, G, about why her paper model does not work. All of the language is quite inexplicit as both of their attentions are focused on the paper model throughout the conversation. [I have tried to give some feel for the overlapping nature of the conversation and also the relative duration of pauses. '...' means a short pause (up to half a second), longer pauses are marked with approximate durations.]

DC: This should fold up there.

G: No, it doesn't

DC: But it don't, does it.

DC: Why, why doesn't it?

G: That should be tilted that way.

DC: That one, *that* way. *[checking] And then?*

G: That way, and ... no, no [turns over piece of paper a couple of times. trying to see how it should go].

Pause of several seconds

DC: laughs

G: [laughs slightly ruefully] If I tilt them top three and then that bit ...

DC: Right.

G: ...would that work?

Pause (2 secs)

(**)

DC: What do you mean "would that work?"? You're asking me. What do you think? You think I'm going to tell you?

G: No [laughs].

DC: No, all right then. [DC takes model]

Pause of seven seconds, both looking at model

DC: [pointing at one particular fold] Is that alright?

(**)

G: Mm-hmm [affirmative]

DC: Now what's the problem now - with that one?

G: That - that should be onto there.

DC: Yup - but it doesn't does it - you get a gap. OK, so what have you got to do?

G: Move that one to there.

DC: OK, you can try that. ... What happens when I fold that one round ... there - does that work?

G: Mm-hmm. No, it isn't ...
DC: So what's what's
G: It's wrong. It's short there and longer there.

DC: Ah yes, that's a length thing, isn't it ? It sort of does. If that were the right length, that would be OK. When I fold this one round, what happens then? Where should it go to?

G: Onto there.

DC: It should go round there, round the back and onto there but it doesn't, does it ? So what have you got to do to that one?

G: That one fixed to that one then.

DC: OK, right. Well, you try that one then. You try that one again. But before you try that, if you hang on a minute, there's some dotty paper which is easier, so you know your sides are all the same length, and try it then.

Teacher leaves.

I am very interested in the exchange marked between the sets of (**). David has not answered her question, in keeping with his desire to have pupils validate their work when they are able to. He does this by drawing her attention both to the question she has asked (by repeating it) and then by asking her whether she thinks it likely that he is going to answer it.

One interesting question is once you have raised the discussion to a meta-level, how do you then return it to the normal one where the form of utterances is not uppermost as the topic of attention. In this instance, David achieved this by means of a long pause and then offering a vary particular focus of attention back on one fold of the model and the question "does this work?". Interestingly, the pupil did not turn the teaching gambit on its head by mirroring his earlier response and replying "What do you think? You think I am going to tell you?". This provides a nice instance of different rules applying to the teacher and the pupils - making meta-comments on the discourse is the prerogative of the teacher.

Moves to and from a meta-level need to be handled smoothly if they are not to be too disruptive of attention and focus, yet not so smooth as to pass unnoticed, as they are often important sites for teaching and, hence, potential learning.

Excerpt 2:

The class has been working on an imaginary number line running round the class and Dave Hewitt has a metre ruler with which he makes loud taps. A move to the left subtracts one from wherever you are and a move to the right adds one. Everything is relative to where you start, which is announced at the beginning of each turn. The class is working on different ways of getting from 15 to 16, and pupils are coming out to the front, tapping out a sequence of moves and then the class announces and records the result. When we enter the extract, the board is as follows:

$$15 + 3 - 2 = 16$$
$$15 + 1 = 16$$
$$15 - 2 + 3 = 16$$
$$15 - 4 + 5 = 16$$

Dave: Is that all right?

[Chorus of yesses.]

Dave: Uh-huh. A different way?

[The boy taps two right and one left, but says nothing.]

Dave: Say again? [Class and he laughs when he realises what he has said.] Start again. Sorry, I've just realised 'say it again' isn't right, is it. That's 15, was it? [Referring to the starting point that was unannounced.]

Class: Add two, take one.

[Teacher and pupil at the board exchange stick and chalk and pupil writes 15 + 2 - 1 = 16 on the board.]

Dave: Uh-huh. Thank you very much. I've got another one.
[Writes 15 + 3 - 2 + 5 - 1 - 3 = 16]
But I'm not too sure whether I have done it right or not. Have I?

Pupil discussion: yes, no, no [more nos]
Dave: It's not quite right - OK, could someone make a change so that it is right.

Thanks, Zena.

(**)

Zena: Can I just rub it out?

Dave: Yes, do [with slight irony, as she has already rubbed out the final 3 with her finger and changed it to a 4]. You can even use a board rubber if you want to.

Zena: [Looks at Dave who is standing at the back of the class] Is that all right?

Pause (2 secs)

Dave: Zena asked a question.

[Chorus of yesses.]

(**)

Dave: Could someone do that for me?

Pupil: [Speaking as he taps] That's 15, 16, 17 18, ...

Dave: [interrupts] I'll tell you what [second thoughts, and puts hand over his mouth.] Sorry, I shouldn't have opened my mouth. [Pupil finishes tapping out sequence.] Right, I shouldn't have opened my mouth. Tell you what, you can concentrate on doing it and we'll do the arithmetic - alright?

Again I am particularly interested in the sequence marked by (**). Zena asked two questions that she may not have distinguished, yet which received quite different treatment from the teacher. The first one was a request for confirmation of procedure and permission - which was given - and the second was a request for verification, which evoked a meta-comment to the effect that she had asked a question.

What sense can be made of the teacher commenting on something that everyone knows who was attending to Zena at the front of the class? Dave had ostensibly made a response, taken a turn in the conversation as the question had been addressed to him. But what he chose to do was to draw the class's attention to the form of the utterance as his response to it, rather than responding to it on the level at which it was made. Once again, the teacher himself had used those exact words a few lines earlier, indeed such a request for audience confirmation was built into the activity. (Not dissimilar to the way a counsellor or therapist might achieve a similar end, namely encouraging reflection on the nature of the speech act just made.)

Eric Love has commented that he often asks: "What do you want me to tell you?" in response to a particular request from a pupil for help, sometimes going further to indicate a range of possible things he was willing to answer (I could tell you exactly the answer, though that might not help you next time you have a similar problem, I could show you how to do this one, show you another one like it, ...). In so doing, he has drawn attention to the fact of the situation as he sees it - that *I* (as teacher) am here to help *you* (as pupil) to learn *mathematics* - but that help might take a variety of forms.

This returning choice to the pupils may seem a preferable solution to the problem of apparently going against direct requests for information or help, which if they are not going to be responded to exactly at face value, require some means or gambit for their deflection. This presents another teaching tension, allied to the didactic tension referred to earlier. If the teacher only makes meta-remarks, only answers questions with questions, the pupil may well lose confidence or trust in the teacher as a source of a conversation about the *content* they are struggling with, the teacher not being straightforward with me. Yet, if the teacher only engages with the content, there is the difficulty of 'teacher lust' (to use Mary Boole's evocative term), of the desire to tell the pupil things, which in part rescinds the possibility of the teacher teaching.

CONCLUSION

Reporting back can place some quite sophisticated linguistic demands on the pupils in terms of communicative competence - that is, knowing how to use language to communicate in certain circumstances. Here, it includes how to choose what to say, taking into account what you know and what you believe your audience knows. Stubbs claims (1980, p. 115): "A general principle in teaching any kind of communicative competence, spoken or written, is that the speaking, listening, writing or reading should have some genuine communicative purpose". Yet this is at odds with my view of the classroom being an avowedly un-natural, artificial setting, in which the structure and organisation of the discourse by the teacher has some quite unusual features.

Pupils learning mathematics in school in part are attempting to acquire communicative competence in mathematical language, and classroom activities can be usefully examined from this perspective in order to see what opportunities they are offering pupils for learning. Teachers cannot make pupils learn - at best, they can provide well-thought out situations which provide opportunities for pupils to engage with mathematical ideas and develop skills in using spoken and written language to that end.

Teachers, in order to teach, need to acquire linguistic strategies (which I earlier called 'gambits') in order to direct pupil attention to salient aspects of the discourse - or indeed the nature of that discourse - while still remaining in 'normal' communication with the pupil. In focusing on instances of meta-commenting, I have tried to highlight specific features of teaching discourse which seem to me central to the teaching enterprise as a whole.

Note: I am most grateful to Barbara Jaworski and Eric Love for conversations on parts of this paper.

References

HMSO	(1982)	*Mathematics Counts* (The Cockcroft Report). London: HMSO.
Jaworski, B.	(1985)	A poster lesson. *Mathematics Teaching*, 113, pp. 4-5.
Love E.	(1988)	Evaluating mathematical activity. In Pimm D. (ed.) *Mathematics, Teachers and Children.* Sevenoaks: Hodder & Stoughton.
Mason, J.	(1988)	What to do when you are stuck. *ME234 Using Mathematical Thinking*, Unit 3, Open University Press.
Pimm, D. and Mason J.	(1986)	*Discussion in the Mathematics Classroom.* Milton Keynes: Open University Press.
Pimm, D.	(1987)	*Speaking Mathematically: Communication in Mathematics Classrooms.* London: Routledge.
Richardson, E.	(1967)	*The Environment of Education.* London: Macmillan.
Stubbs, M.	(1975)	Organizing classroom talk. Occasional paper 19, Centre for Research in the Educational Sciences, University of Edinburgh.
Stubbs, M.	(1980)	*Language and Literacy.* London: Routledge and Kegan Paul.
Stubbs, M.	(1983)	*Language, Schools and Classrooms.* London: Methuen.

'MATHOPHOBIA', PYTHAGORAS AND ROLLER-SKATING

Richard Winter

What is it about mathematics that makes it seem so 'difficult' and remote? Why is it that progressively-minded educators seem to have been relatively unable to make the learning of mathematics 'enjoyable', despite some moderate success with, say, the learning of science and history? At the heart of my argument in this essay lies a distinction between 'work' (activity governed and limited by rules imposed by others) and 'play' (where a grasp of the basic situation is the beginning of creative individual improvisation). All successful education, I would argue, aspires to the condition of play.

So why does maths, in particular, always feel like work, like 'doing as you are told', and not like play, like, say roller-skating? The root of the problem lies with Pythagoras, with the ancient notion of mathematics as the revelation of the Divine Forms which supposedly determine the world of our experience. In other words, the problem with mathematics (and the reason for the anxieties of 'mathophobia') is a problem of mathematics as a cultural form - a combination of mystery and power: maths educators are still haunted, I shall try to show, by the repressive basis of the tradition they are trying to liberalize.

The use of the term itself ('mathophobia') may be seen as an attempt to dignify the panic with which so many of us react to formal mathematical education, and at the same time to convert a cultural issue into an individualized shortcoming. For example, Resek and Rupley (1980) claim the 'Mathophobia ... is an excessive ... irrational and impeditive dread of mathematics', whose major symptom is a rigid fixation upon 'isolated rules', in contrast to a healthy 'understanding' of 'a collection of interrelated concepts' (pp. 423-4). For politicians and employers, all this means the 'failure' of maths education, and complaints about 'low standards' of numeracy in the future workforce; and this, in turn, adds to the sense, on the part of progressive maths educators, that they face an uphill struggle against wide-spread negative attitudes. Thus, Kathleen Hart's foreword to her well-known book Children's Understanding of Mathematics 11 - 16 (1981) ends, almost plaintively, with only a mere 'hope' that 'mathematics can become more relevant, attainable, and even [!] friendly to ... school children'.

Hart's work is a serious attempt to clarify the nature of the separation between 'formal' mathematics and common-sense, experiential reasoning processes, and much progressive writing on maths education is explicitly concerned with pre-cisely this issue. However, this is not a recent insight but a long-standing unsolved problem. One of the most widely influential texts is the Cockcroft Report (1982), and it is significant that Cockcroft begins his chapter on primary maths by quoting an HMSO publication from as long ago as 1937.

First by way of introduction, should come practical and oral work designed to give meaning to, and create interest in, the new arithmetical concept - through deriving it from the child's own experience - and to give him confidence in dealing with it. (Cockcroft, 1982, p. 83)

What offer of a 'friendly relevant attainability' could be more explicit? Why has it, apparently, been ignored for fifty years? Or what can have defeated its intentions? Cockcroft straight away gives a clue:

> The primary maths curriculum should enrich chil-
> dren's aesthetic and linguistic experience, provide
> them with the means of exploring their environment,
> and develop their powers of logical thought. (p. 84)

Immediately we notice that the process of enrichment seems to be one-way, and that children here seem to lack a means for exploring their environment until it has been 'provided'. We may also ask what powers of logical thought they have already, and thus what starting-point is to be taken for 'development'. Cockcroft has anticipated our question:

> To speak of logic in connection with young children
> may surprise some people, but no highly theoretical
> notions are involved. It is rather a matter of describing
> things accurately, noticing their resemblances and
> differences, and saying how they are related to one
> another ... In its most straightforward forms, the
> activity of sorting objects and of recording results in
> diagrammatic form is practised in most infant classes
> and forms the basis on which the concept of number is
> built. As children become older, it can develop into the
> more sophisticated activity of sorting shapes which
> vary in colour, size, and thickness according to their
> attributes such as, for example, 'large and blue'. (p. 86)

But, in attempting to reassure 'some people', Cockcroft inadvertently lends support to other people's worst suspicions; that mathematics educators suppose that young children's cognitive processes begin with 'perception' (whose task is 'accuracy') and that 'concepts' must then be 'built' (by educators) through later, 'more sophisticated' activity. Such suspicions are confirmed when we read that the elements of mathematics education are: Facts, Skills, Concepts and General Strategies, listed in that order both in Cockcroft (p. 71) and Mathematics 5-16 (HM Inspectorate, 1985, p. 7).

The strong implication here is that the relationship between experience and analytical thought may be conceived as a progression, from fact to concept, from concrete to abstract, and from simple to complex. Ormell (1969) has described this as 'the doctrine of Logical Sequence' (p. 50) and as the 'ideology' of school mathematics. It is a classificatory logic which creates a hierarchy from 'sense data' or 'intuition' (as a naïve starting-point) to 'abstraction' and 'conceptualization' (as the achievement of tutored reflection). It is this distinction which, in the most 'unfriendly' fashion, introduces a devaluation of children's spontaneous thought processes and appears to render analytical thought on their part as a remote and barely attainable goal. And it is this distinction which the research reported here is intended to subvert. (The report is based partly on a series of observational notes on the development of my daughter Jessie, and partly on transcribed tape recordings of maths work in schools with children of various ages.)

SIMPLY 'PERCEIVING FACTS'

At what age do children 'start' by simply 'perceiving facts'? The following comment from Trevarthen (1982) refers to babies aged from two months onwards:

When a mother teases an infant, she makes a joke out of the feelings, actions, or interests of her baby, gently thwarting or negating what the baby wants or expects. That this is a source of pleasure indicated by laughter, is proof that the baby is interested in the clash or mingling of intentions. (p. 99).

If at two months old one is enjoying the ambiguity of experience, it is hard to imagine when consciousness could ever have 'started' as a mere retina for the registration of 'facts'. Consider, for example, the following observation, made when Jessie was a mere ten and a half months:

> Jessie was holding a bottle of baby milk in one hand, and in her other hand she was holding a long tapering plastic cone from a set of toy pyramid rings. Switching them from hand to hand, she seemed to have a moment of confusion and put the cone into her mouth instead of the bottle. She immediately took it out, looked at it, and put the bottle into her mouth instead. She then switched back, putting the plastic cone into her mouth, took it out, and laughed broadly.

The joke here seems to be the way in which the vagaries of experience continually elude the simple categories into which we think we have got life sorted. Jessie happily enjoys the always ambiguous relationship between concepts and experience. As David Bloor says: 'The transition or link between arithmetic and the world is the link of metaphorical identification between initially dissimilar objects. This is the key to the general problem of the wide applicability of arithmetic' (Bloor, 1976, p.92).

If infants can take their pleasures in such philosophical ways, one wonders indeed what can have happened to the following eleven-year-olds:

> When they found that [their] predictions were not confirmed by the data, many pupils reverted to past experience, hunches, or cynicisms ... pupils became confused, either clinging to what they believe must be the mathematically correct answer, despite contrary

> results, or abandoning theory altogether. (Kapadia,
> 1984, p. 47, quoting the APU document on mathemati-
> cal development)

Bloor's reference to the metaphoricity of concepts contrasts interestingly with Cockcroft's characterization of mathematics as an 'unambiguous' language (Cockcroft, 1982, p. 1), and serves to introduce an example of spontaneous infant mathematics. Jessie was aged two years and seven months at the time of the following incident:

> At this time we regularly played dominoes - the
> coloured variety - and always began by taking seven
> each, withdrawing them one by one from the pile in
> parallel with each other, intoning the numbers one to
> seven. Today, at breakfast time, there was a pile of
> chopped dates. Previously, when I had said, 'Take
> some and leave some for me,' she had taken all the
> pieces but one. On the day in question I said, 'You
> have half and leave me half; that means you have the
> same amount as me, otherwise it's not fair.' She
> immediately began to separate out two piles, saying:
> 'For you, for me; for you, for me ...' In the middle of
> this operation she pointed to a large piece of date and
> said, 'That's a double-six.'

The next day I said, 'Shall we share out the dates again?' and Jessie said, 'Yes, and shall we do it taking turns again?' This latter phrase stems from my interventions between her and her friend when both wished simultaneously to play with the same toy, that is, 'Take turns to play with it.'

The significance of this episode seems to me as follows. Jessie spontaneously used her experience of counting dominoes as a technique for solving the problem of making an equal division of the dates. This linkage suggests a full grasp of the notion of mapping one set on to another in order to create two equal sets. Calling a large piece of date a 'double-six' makes clear that, for her, problem-solving in mathematics was a form of metaphorical thinking. A further metaphor is created when she makes the link with 'taking turns', showing how a range of potentially relevant concepts and strategies is being assembled. Jessie had received no specifically mathematical 'teaching' up till then.

Here, to conclude this section, is an incident which seems finally to show the inadequacy of Cockcroft's suggestion that a pre-school child's conceptual thought may be thought of, typically, as an ability to sort objects. Jessie at this time was four and a half years old.

> *Jessie: (Handing me her cup of cool Ribena) Can I have*
> *warm Ribena, please?*
>
> *RW: Yes but you'll have to wait until the kettle boils (I*
> *had just turned it on); otherwise (pause, as I thought hard*
> *how to put it simply) it won't be warm enough.*

Jessie: Otherwise (pause, then slowly), there will be too
much water and not enough hotness.

To put this down as 'intuition' would indeed be a put-down. Let us rather appreciate it as a skilful and successful piece of analytical thinking, in which an abstract category is created in order to organize the significance of a range of experiences in the light of a hypothetically conceived prediction.

LEVELS OF DIFFICULTY

Hart (1981), in her book on secondary mathematics, is concerned throughout to argue that many of the tasks which mathematics teachers set for pupils aged eleven to sixteen are 'very difficult', and she arranges these tasks in a 'hierarchy' which represents differing 'levels of understanding', ascending from Stage 1 to Stage 4 (p. 187). She describes this hierarchy as follows:

> Stage 1 items were mainly concerned with the under-standing of the meaning of new conventions, whereas Stage 2 seems to be concerned with the application of these conventions and is therefore more concerned with the understanding of when they should be applied than with the pure knowledge of the language of symbols. What was introduced in Stage 1 becomes operative in Stage 2 ... The main difference between the items in Stage 2 and those in Stage 3 in the appearance of the first measure of abstraction - questions are not always tied to a diagram, or the child is asked to hypothesize about situations which are not shown ... Stage 4: Items in this last stage involve abstraction as well as the application of a fund of knowledge to the solution of problems ... This is the type of mathematics recognised by the professional mathematician. (pp. 194, 199, 203, 204)

It is clear that we have here another version of the two familiar hierarchical sequences: 'Fact-Concept-Strategy' and 'Concrete-Abstract'. It also evokes another, equally interesting sequence, namely, 'Teacher exposition - Discussion - Practical work - Routine practice of skills - Problem-solving - Investigation'. This sequence is found both in the Cockcroft Report (p. 71ff.) and in Mathematics 5-16 (HM Inspectorate, 1985, pp. 38-42). It is not explicitly presented as a pedagogical sequence, but it is the powerful subtext, whereas many teachers, especially of subjects other than maths, would wish to argue for the validity of other sequences of these elements, or even for an exact reversal of the series.

The following examples, I think, help to raise crucial questions about the assumptions behind Hart's account of these hierarchical stages. Here, to begin with, is an account of a session with reception infants.

I made a 'spinner' game to play with a group. (There were six children, supposedly 'able but lacking in concentration'.) The spinners had the following forms:

The game was that two children each had a pile of counters; if the spinner landed on a one the child received a counter from its partner; if it landed on a two the child had to give a counter to its partner. At first the children played on their own. Then:

RW: Right, who hasn't had a go yet? James and Anna? OK, look at the spinners very carefully. Remember, you win if you get a one. Look carefully at all the spinners and see if you can choose the one where you think you are going to win most counters. OK, after you've taken it, just take it out. Right, James, you chose that one (X) and Anna, you chose that one (XX). Now, the rest of you have a look at the two spinners that James and Anna have taken. Now, who do you think - look very carefully - show them the spinners so the others can see - who do you think is more likely to win, James or Anna?

(All agreed that James would win.)

Why do you think James is going to win? Yes, Karen?

Karen: *Because it's got six ones and one two.*

RW: And what about Anna? Yes, Phillipa?

Phillipa: *She's got five ones and one two.*

RW: Actually ... Count those again. I don't think it's right.

(It was established that James's spinner had seven ones and Anna's had six ones.)

RW: And that means that James is going to win?

All: Yes. (Confident and eager tone, not just a ritual chorus.)

I conclude from this that five-year-olds, who have had no 'teaching' on 'ratio' or 'probability', and who still make slips in counting up to seven, can 'understand' the probability implications of ratios such as 1:2, 1:3, 1:5, 1:7, 2:2 and 3:6. They did not need Hart's Stage 1 'conventions'; they understood that a certain aspect of numerical relationships 'should be applied' (Stage 2), and this involved 'abstraction' and making hypotheses about situations that were not shown (Stage 3) and 'the application of a fund of knowledge to the solution of problems' (Stage 4). If indeed this is 'the type of mathematics recognised by the professional mathematician', then some children arrive at school able to do it.

I then decided to explore further a particular example of 'Stage 4' difficulty, namely the question 'How many different numbers could you write down which lie between 0.41 and 0.42?' (Hart, 1981, p. 205). Note that this is presented as the most difficult level of question expected of children aged eleven to sixteen. The following conversation took place between myself and Jessie, aged eight and a half:

RW: If I give you that piece of paper (A4 size) and a pair of scissors, how many pieces do you think you could cut it into?

Jessie: Millions.

RW: Millions?

Jessie: Well, I don't mean millions, I mean lots.

RW: If you had a very, very, er, almost a magic pair of scissors, so that you could cut things as small as they could possibly be, then how many would it be?

Jessie: Er, about a hundred or a thousand.

RW: About a hundred or a thousand. If you had a magic pair of scissors, though, you could cut it as small as it could possibly be.

Jessie: A million, a billion.

RW: Right: if you imagine each of those small pieces of paper that you had cut it into, right, and you had this magic pair of scissors that can cut things as small as they go, then could you cut any of those small pieces of paper?

Jessie: You might.

RW: You might?

Jessie: It depends, 'cause if they're as small as they can go ... that's why they might. You might, or you might not.

RW: When might you, and when might you not?

Jessie: It depends how small 'as small as they can go' is.

RW: Hmm. Well, how small is 'as small as you can go'?

Jessie: Nothing.

RW: Each one would be nothing. So how many do you think there would be?

Jessie: Millions, billions, trillions.

RW: Right, if I gave you this other piece of paper, which is much smaller, right, and you did the same, and you cut that up as small as you could, in the same way, how many little pieces would you have at the end?

Jessie: About the same, because it would be cut up into nothing, and it would be the same.

RW: Great. That's the first question. Now then: if I start writing numbers like this, 1, 2, 3, 4 ... and we were to play a game, and you were to start writing numbers (gives Jessie a piece of paper and a pen), if we had a competition to see who could write the biggest number ... suppose I write that number, what's that?

Jessie: A hundred.

RW: And what's that?

Jessie: A thousand.

RW: If I write a thousand thousand, how many is that, do

> *you know what that's called?*
>
> *Jessie: A million?*
>
> *RW: Right, is that the biggest number you can write, a million?*
>
> *Jessie: No*
>
> *RW: No. Right. Well, how big? What I'm asking is: how big is the biggest number you can write?*
>
> *Jessie: You can't make a biggest number.*
>
> *RW: Why not?*
>
> *Jessie: Because you can just go on doing one, two, three, dot, one two three ...*
>
> *RW: Now let's just do another one. Suppose I have a line, and I wanted to divide it into little bits, how big would the smallest little bit be that you could divide that line into?*
>
> *Jessie: The smallest bit?*
>
> *RW: Hmm.*
>
> *Jessie: Nothing.*
>
> *RW: So if I asked you how many points are there in that line.*
>
> *Jessie: What does that mean?*
>
> *RW: Well, the 'point' is that little nothing-bit that you said. And suppose I said, 'How many "points" are there in that line from there to there?'*
>
> *Jessie: How many nothings?*
>
> *RW: Yes.*
>
> *Jessie: As much as you want them to be.*
>
> *RW: OK. Now, if I ask you the same question about that line (much shorter)?*
>
> *Jessie: It would be the same.*

Jessie did not know what a 'point' meant in this context, and, I discovered afterwards, had not met the word 'infinity'. Yet, despite lacking many of the agreed conventions, an eight-year-old could grasp the essential conceptual problem set by a Stage 4 question, and solve it. What is 'difficult' about this question, I would argue, is not its intellectual sophistication but its format, which is unhelpful to the point of being misleading. Its only clue is the emphasis: 'How many numbers could you write?' To cope intelligently with such a format would either require an enormous degree of combative confidence in the face of authority or the contingent possession of that particular 'isolated rule', but that would take us back - ironically - to the supposed symptoms of 'mathophobia'! (See my second paragraph on p. 82).

PROGRESS

Hart concludes from her investigation that 'Mathematics is a very difficult subject for most children. We have shown that understanding improves only slightly as the child gets older' (Hart, 1981, p. 209).

The spinners game was tried with a group of first-year secondary children. The spinners were as follows:

> The children were asked to choose a spinner for them-selves and one for their partner, in order to maximize their chances of winning. Philip's response was inter-esting, and not untypical of most of the group, who were roughly seen as rather below average for the school. Philip initially chose C for himself and B for his partner, and explained his choice afterwards by say-ing, 'It spins well,' which he could not possibly have known at the moment he chose it. Then, after a pause:

> *Philip: B has a lot of Xs and therefore a lot of chances.*
> *RW: Chances to what?*
> *Philip: Chances to spin an X.*

He then went on (unprompted) to count the Xs and to conclude that he should have chosen B for himself (correct) and given his partner A (incorrect).

Philip here shows a relatively insecure grasp of the principle needed. He is initially impatient, and in the end he does not follow his insight through.

How comes it that a group of eleven-year-olds can find difficulty in a task which is successfully tackled by a group of five-year-olds just entering school? To put it down to a predictable 'seven-year difference' (Cockcroft, 1982, p. 100) begs the question: a difference in what? It is as though for Philip and his friends six years of maths teaching has not only failed to accomplish any progress, but has tended to destroy an ability originally present. An article by Fischbein and Hazit (1984 seems (inadvertently, and against the authors' explicit intentions and conclusions) to offer evidence of precisely this destructive process at work. Having mounted an experimental teaching programme for one of two matched groups of children, the clearest result shown is that the proportion of pupils offering no answer to the post-text questions was generally higher in the group which had received the teaching than in the control group, especially, and dramatically, in the lower age-group (ten year-olds) (Fischbein and Hazit, 1984, pp. 14-19).

This seems to suggest that children's ability to carry out maths tasks can easily be undermined. A group of six-year-olds with whom I attempted to play the spinners game as a complete stranger responded by (a) choosing the 'wrong' spinner, and (b) justifying their choice by saying variously that the one they had chosen would win because it was 'bigger' or because it was 'smaller'. Jessie, who has largely featured as the heroine of this text, when faced at the age of eight with two spinners bearing a tick and dots in the ratio of 1:5 and 1:7 (the tick twins) chose the 'correct' spinner (1:5 but explained:

> **Jessie:** 'Cause it's got less dots and it's got flatter sides; that (other) one's got more chance of rolling.
> **RW:** Is it more important that it's got more dots or that it's got flatter sides?
> **Jessie:** I don't know. (She scans my face for clues.)

Flatter sides, I think.

My interpretation of this is that a crucial element in her response was that we had just had an emotionally fraught argument. Having played briefly with the spinners and - presumably - dissipated the distracting emotion, she then responded quite skilfully to a choice between a spinner with three ticks and five dots and another with four ticks and twelve dots: she correctly chose the 3:5 ratio as more favourable, and explained: "'Cause that one, it's got three ticks and it's only got five dots, but that one's only got four, er, four ticks and it's got lots of dots.'

In other words the emotional tone of the interactional setting in which the mathematical task is set can either help or hinder, support or distract. The problem with 'teaching' is that it can seem to the learner more like an infliction than a gift, as the following conversation seems to illustrate. (Jessie is five years old at this point, a reception infant.)

Every morning, I give Jessie her dinner money, that is, 35p. About a week ago, instead of giving her the exact money, I emptied my purse on to the table and said, "Take thirty-five pence from that.' This she did.

Today, before I had mentioned dinner money, Jessie asked, 'Dad, please will you put the money out, so I can choose it?'

During the 'choosing' of the money Jessie monologued: 'Two tens is twenty. And ten. And five. How much does that make? Thirty-five. Right, I'll put that in my purse.'

When Jessie showed me her handful of money, she had two tens and three fives. So when she had said 'And ten' she must have collected two fives.

I asked, 'What is the opposite of choosing?'

Jessie said, 'Working ... except that you chose to work.'

'Who else chooses what work to do?'

'Miss C (her teacher at school).'

At the end of this she was in an excellent mood and volunteered to put the remainder of the money back in my purse.

Counting, here, for Jessie, was an opportunity for the exercise of autonomy. The number system is used as a powerful skill, enabling a situation to be controlled. Exercising one's ability to count, I conclude, can be as inherently enjoyable, as self-enhancing, as exercising one's ability to roller-skate or turn cartwheels. The result of doing it is a sense of self-confidence; hence Jessie's offer to help me at the end of her counting; she had generated a surplus of energy she could generously distribute. Counting = choosing = the opposite of being told what to do (being given money). Counting, here, is thus the opposite of work. Work is what adults (parents and teachers) choose to do for themselves, and what they tell children to do.

WORK OR PLAY?

Mathematics could be like roller-skating, but usually it is like being told to stop roller-skating and come in and tidy your room. This is not a superficial matter. The failure of maths to be 'friendly' to school children is due to a fundamental 'unfriendliness' in the epistemology which still seems to underlie the thinking of maths educators, and which Cockcroft and Hart exemplify even as they lament its consequences. Subconsciously, I would argue, mathematicians are still haunted by the ghost of Pythagoras, for whom mathematical knowledge was a divine revelation.

Although Cockcroft (1982) invokes 'practical methods' and 'the child's own experience' from the very beginning of the chapter on primary maths (p. 83), his general emphasis on mathematics as a language for the ordering of experience is continually balanced against reminders that mathematics can or should also be enjoyed 'for itself' (p. 2), as a 'doorway' to a magic work (p. 3). Inside every Piagetian maths educator wishing to encourage children to derive concepts from the structure of action (see Winter and Lang, 1983) lurks a Pythagorean educator wishing to reveal to children the eternal Divine Forms of which children's experience must inevitably be but a confused anticipation or a pale reflection.

As adults, we all have difficulty in treating children's thought processes as other than inadequate gropings towards our own assumed achievements, but for most of us this is a sociological problem concerning the status and power relations of adult-child interactions. The particular unfriendliness of mathematicians' ways of devaluing children's experiences, however, is that in mathematics the power of the adult over the child is reinforced by the distant cultural echo of an ancient theological justification. At best the maths teacher approaches the child's experience as a benevolent theocratic despot. The difficult format of mathematical tasks (which, I have suggested, does not necessarily represent a particularly difficult conceptual demand) confronts the child as a divine wrath. The terror of the fifteen-year-old confronted by 'How many numbers are there between 0.41 and 0.42?' is the terror of the citizenry confronted by the pronouncements of the Delphic oracle, whose practical implications, wrapped in semantic mystery, were always only understood by the time it was 'too late'. And it is at least questionable whether most maths teachers, as priests of the mathematics temple, are unambiguously committed to liberating youthful citizens from their sense of awed subjection to its mystery.

Certainly, it is hard to understand how Cockcroft, in a report which is in many ways committed to deriving mathematical education from the learner's experience, can say: 'We do not discuss the mathematical development of children of pre-school age' (Cockcroft, 1982, p. 83). And this four years after the appearance of Margaret Donaldson's celebrated work on the capacity of pre-school children for deductive reasoning (Donaldson, 1978, p. 58). Instead, borrowing from the well-known work of Abraham Kaplan, I would propose that the thinking of young children is not to be characterized as 'intuition' (with its implications of pre-logical, pre-conceptual, pre-rational) but as the respectable operation of 'logic-in-use' (Kaplan, 1964, p. 13ff.) which characterizes any process of intellectual discovery.

Mathematics, in contrast, always manages to imply that a respectable logic is precisely what children don't possess until it has been provided for them. (In the end this is even true of Hart's account of 'idiosyncratic schemas'.) The maths lesson may start with games and constructions, but children know that at some point it will suddenly turn its other face and say, 'Stop roller-skating, it's time to come in and work.' Until mathematics educators can fully exorcize the ghost of Pythagoras, mathematics learners will continue to feel that their precious experience is being subjected to a covert act of authoritarian expropriation.

In this respect, 'the ghost of Pythagoras' in maths education may be taken as a metaphor for the general process of mystifying the basis for cultural domination. Mathematicians complacently assume that the problem of 'mathophobia' is due to the inherent 'difficulty' of their subject - its 'Divine' abstraction. Yet this should rather be seen as the self-interested avoidance of a crucial cultural and political issue: the hierarchical separation of 'mental' from 'manual' labour. This widespread assumption denies the intellectual accomplishment implicit in all competent activity, and thereby reinforces the superiority of those who handle symbols for a living, like bankers, bureaucrats, systems analysts, and - in spite of their apparent good intentions - maths educators.

ACKNOWLEDGEMENTS

The work reported here was supported by a bursary from the Research Committee of the Anglia Higher Education College. Many of the ideas originated in affectionate controversy with my colleague Peter Ruane and with the late Bryan Lang, to whom I am deeply grateful.

REFERENCES

Bloor, D.   (1976)   *Knowledge and Social Imagery.*
                                London: Routledge.

Cockcroft Report(1982)   *Mathematics Counts:* Report of the Committee of In-
                                quiry into the Teaching of Mathematics in Schools
                                under the Chairmanship of Dr W. H. Cockcroft.
                                London: HMSO.

Donaldson, M. (1978)   *Children's Minds.* Glasgow: Fontana.

Fischbein, E.
and Hazit, A.   (1984)   'Does the teaching of probability improve probabilistic
                                intuitions?', *Educational Studies in Mathematics* 15.

Hart, K.   (1981)   *Children's Understanding of Mathematics 11-16.*
                                London: Murray.

HM Inspectorate(1985)   *Mathematics 5-16.* London: HMSO.

Kapadia, R.   (1984)   'Probability - the subjective approach', *Mathematics
                                Teaching* no. 108.

Kaplan, A.   (1964)   *The Conduct of Inquiry.* San Francisco: Chandler.

Ormell, C.   (1969)   'Ideology and the reform of school mathematics',
                                Proceedings of the *Philosophy of Education Society of
                                Great Britain III.*

Resek, D.
and Rupley, W. (1980)   'Combating "Mathophobia" with a conceptual ap-
                                proach toward mathematics', *Educational Studies in
                                Mathematics II.*

Trevarthen, C.   (1982)   *'The primary motives of cooperative understanding',* in
                                G. Butter worth and P. Light, eds Social Cognition.
                                Brighton: Harvester.

Winter, R.
and Lang, B.   (1983)   'Mathematics teaching - a one-ended bridge?', *Journal
                                of Curriculum Studies* 15(4).

PGCE STUDENTS AND INVESTIGATIONAL APPROACHES IN SECONDARY MATHS

Barry Cooper

ABSTRACT

Several researchers have recently drawn attention to aspects of the teacher shortage in maths: both its causes (Smithers and Hill, 1989) and its consequences (Straker, 1987). In particular, there has been considerable concern about the quality and quantity of entrants to Post Graduate Certificate in Education courses (Hayter, 1989). It is within this context that a move has occurred within secondary schools towards a greater use of investigative approaches in mathematics (HMSO, 1987).

This paper reports on one aspect of a qualitative study of a university cohort of PGCE mathematics students: their responses to investigational work in their teaching practice schools. These students had a variety of backgrounds. Some were recent graduates in maths or related disciplines, while others had taken their degrees nearly twenty years ago. All had experienced a largely algorithmic approach to maths teaching at school. Several had suffered the fairly typical experience for those who study maths at university of being *warmed up* at school and then being *cooled out* at university.

The paper briefly discusses the investigational approach. It then describes the school and undergraduate experience of the student group, in order to show that their university careers had resulted in many of them lacking confidence in themselves as "mathematicians" (Bibby, 1985). It is suggested that, when faced with the new uncertainty represented by "investigations", as well as the usual problems of classroom control experienced by PGCE students (Lacey, 1977), they might have been expected to have favoured the algorithmic approach to maths they had experienced, and largely succeeded with, at school.

The main body of the paper presents detailed case studies of four students, two of whom were placed in a very pro-investigational school. The responses of these four to investigational approaches are described as they developed during the year. It is shown how, in some cases, a lack of confidence, either about undertaking less didactic classroom roles or about the mathematical content implicit in many investigations, would seem to account for students' initial responses, some of which were characterized by high levels of anxiety and confusion. As some students' confidence grew, and as they perceived the possible motivational advantages of an investigational approach, they began to be more positive about such methods. In other cases there would also seem to have been a value preference at work, and some students remained unhappy about various aspects of investigational work at the end of the course.

The paper ends by summarizing the students' shifts in perspective and briefly considering ways in which the PGCE course itself might make the transition to investigational work more easy and fruitful for students such as those described.

INTRODUCTION
Delamont (1989) has recently pointed to the relative dearth of work on science and maths within the sociology of education. Given the central role of these subjects in the school (now National) curriculum and the key role they play in the production of young people's beliefs about their intellectual capacities, this is perhaps initially surprising. It no doubt needs to be understood in terms of the academic biographies of those doing work in the sub-discipline. This paper reports on one aspect of a study which is intended both to contribute to the development of a sociology of maths education and to be policy-relevant. Using a qualitative approach, it portrays and begins to analyse the responses of university Postgraduate Certificate of Education students to investigational approaches in secondary maths. None of these students had experienced such approaches themselves at school. The study as a whole explores students' perspectives of pedagogy and teaching as a career as they move through a school-based PGCE and into teaching (or other occupations). It is hoped eventually to combine the insights a qualitative study makes possible with those generated from survey style approaches (for example, Smithers and Hill, 1989). Before discussing these students' experiences of maths and their responses to investigations, I shall briefly discuss the investigational approach.

INVESTIGATIONAL APPROACHES IN SCHOOL MATHS
The Cockcroft Report (1982), *Mathematics Counts*, officially legitimated the investigational approach as *one element* of school maths (paragraph 243), and as something which should characterize all maths lessons rather than as necessarily to do with "extensive piece(s) of work" (paragraph 250). While investigations would sometimes take the form of "projects", more generally (paragraph 250):

> The essential condition for work of this kind is that the teacher must be willing to pursue the matter when the pupil asks "could we have done the same thing with three other numbers?" or "what would happen if ...?" ... The essential requirement is that pupils should be encouraged to think in this way ... There should be willingness on the part of the teacher to follow some false trails and not to say at the outset that the trails lead nowhere. Nor should an interesting line of thought be curtailed because "there is no time" or because "it is not in the syllabus".

In 1985, the HMI's *Maths 5 to 16* supported this view. It also provided a description of the variety of activities that might come under the headings of "problem-solving" and "investigative work" - between which "clear distinctions do not exist". Problems might have a unique solution, no solution or several solutions with different merits. Others might have a solution if more information were available (p.41). "In broad terms", it is useful to think of problem-solving as a convergent activity and of investigative work as more divergent (p.42). In an investigative approach, pupils "are encouraged to think of alternative strategies, to consider what would happen if a particular line of action were pursued, or to see whether certain changes would make any difference to the outcome" (p.42).

The 1985 GCS/CSE Criteria included practical and investigational skills (TES, 29/3/85). An investigational approach would involve considerable change for many maths teachers (Cooper, 1976, pp.22-36; HMI, 1980), requiring a shift from *transmission* to *interpretation* views of teaching and learning (Barnes, 1976).

Pedagogical approaches, and their social distribution, have been heavily contested since 1950 (Cooper, 1985a & 1985b; Ruthven, 1986; Mellin-Olsen, 1987; Abraham and Bibby, 1988; Ernest, 1989; Howson, 1989; Noss, 1989). Most recently, during argument about the various reports of the Mathematics Working Group and the National Curriculum Council on maths in the National Curriculum, the New Right (Quicke, 1988; Whitty, 1989) has succeeded in moving discussion towards a concern for "basics". Examples are their success in getting pencil and paper algorithms for long division and multiplication returned to the attainment targets, and the removal of Profile Component Three, which amongst other things, included such objectives as the "ability to co-operate within a group" and "independence of thought and action". All this is recorded in DES (1988; 1989), and National Curriculum Council (1988).

For most of the New Right, maths is a necessarily hierarchical body of knowledge, not a looser network of ideas and practices through which children might find different routes. Coldman and Shepherd (1987) provide one example: the GCSE proposals were "the latest stage in a disastrous process that has seen school maths drift towards becoming a low-level empirical science", whereas "the structure of mathematics demands a formal exposition". The proposals were a threat to a subject that had "traditionally provided a first class training in disciplined and exact thinking".

In the LEAs in which the students in this study carried out their teaching practice, several secondary schools had been working towards a more investigational approach.

THE COHORT

The cohort have come together as a result of processes I can only briefly describe. In 1938 75% of maths graduates entered school teaching (Straker, 1987, p.127). In 1961 about 40% of honours graduates in maths did so (Hayter, 1989, p. 258). In the late seventies 12% of a much larger group of maths graduates entered (Hayter, 1989, p. 258). Since then, opportunities offered in industry and commerce have continued to attract graduates and the 1984 figure was down to 8% (Straker, 1987, p.127). A variety of factors including pay, conditions of service and diminished public prestige have made teaching much less attractive (Smithers and Hill, 1989). Smithers and Hill (1989) have also argued that students of science and maths, particularly males, do not seek the sort of occupational satisfactions that teaching, as a "person-oriented" profession, can offer. It is not surprising therefore to find that entrants into PGCE courses for maths are generally less well qualified than those in some other subject areas. Table 1 shows this clearly.

Table 1: Degree class and PGCE subject group: universities, 1986 entry

(percentages at each level within maths, history and as a whole)

	Higher Degree	1st	2i	2ii	Undivided 2nd	Third	Pass/ General	Other
Overall	3.0	2.9	33.2	39.2	2.5	9.4	7.3	2.5
History	3.3	3.3	47.2	40.7	3.9	0.7	1.0	-
Maths	2.8	4.3	19.3	27.9	5.2	19.7	17.8	3.0

Source: Hayter (1989) from UCET figures for the 1986 entry

The significance of the degree class - what sort of undergraduate experience it indicates - will be discussed later. But Table 2, setting out basic data for the Sussex group, shows that they are, as a whole, a somewhat better qualified group than the total 1986 maths entry to the university sector PGCE (though major differences exist *within* the group). There were seven women to six men. Two men and two women were markedly older than the rest. Table 2 also shows that only eight are going straight into teaching. Only four of this eight will be involved in teaching 11-16 year olds in comprehensive schools. I intend to explore in another paper what led these thirteen students to decide to teach or not.

THE PGCE COURSE

On the Sussex course (Lacey, 1977; Furlong et al, 1988), students spend three days in school and two days in university each week for much of their year. Responsibility for assessing practical teaching is delegated to Tutors in school. Each week in the university the students attend a subject-based Curriculum Group and a Personal Tutor Group, which includes the students from one or two schools and is the main forum for the discussion of educational theory (in principle, in relation to their work in school). The course rhetoric lays great stress on reflection, students having to produce a Course File demonstrating this. The typical Curriculum Group for the maths students studied consisted of a feedback session, during which they exchanged news about work in school, followed by a mathematical investigation in which they played the part of pupils. Their perception was that there was little discussion of the rationale for the investigative approach. It was rather assumed to be a good thing. They practised rather than discussed it. Later in the year there were also sessions involving teachers from the schools on such topics as gender and maths, and marking. There were also sessions, particularly later in the year, orientated more to the traditional issue of how best to teach a topic such as algebra or negative numbers.

Table 2: The student group: basic data

	Age	Own secondary schooling	Degree(s)	Previous work	Post-course destination
Emma	21	Comprehensive + sixth-form college	Maths: 2ii	n/a	Comprehensive, 11-18
Linda	21	Comprehensive	Maths: 2ii	n/a	Privatised utility
Joan	21	Private girls	Maths: 1st	n/a	Banking? Not teaching
Jenny	22	Comprehensive (was secondary modern)	Maths: 2ii	n/a	City? Not teaching
John	24	RC secondary modern + FE + sixth-form college	Maths: pass	Clerk, junior manager	Comprehensive, 11-16
Sandra	21	Comprehensive + FE college	Maths: 2ii	n/a	Chartered accountancy
Neil	26	Boys' grammar	Finance and Accounting: 2ii	In city firm	Chartered accountancy
Martin	46	Grammar	Electrical Engineering: 1st + D.Phil.	Training manager, etc, in computing companies	Sixth-form college
Paul	21	Grammar turning comprehensive	Maths: 1st	n/a	Comprehensive, 11-18
Richard	27	Grammar + sixth-form college	Maths: 1st	Maths research + business analyst	Sixth-form college
Diane	39	Girls' grammar	Sociology: 2i; + OU Maths: Pass (much later)	1 year of social work then housewife	Comprehensive, 11-16
Sheila	38	Girls' grammar	Maths: 2ii	Mix of work in insurance and as housewife	Sixth-form college
Michael	39	Secondary modern + technical college	Computer Studies: 2i (1988)	Computer programmer/ analyst	Girls' Grammar School

THE RESEARCH

The research has been largely interview-based. The thirteen students have been interviewed at least once in each term. Six have been interviewed four times, six five times and one six times. For each student, one interview followed lesson observation by the interviewer. All interviews, bar one, were taped and transcribed. I also observed one Curriculum Tutor session, in November. I occasionally talked to students over coffee. I only related to two in a (personal) tutorial role. The students who are entering teaching next year have all agreed to further interviews and observation.

COHORT'S PREVIOUS EXPERIENCE OF MATHS

This section demonstrates that many students did not have a history of uninterrupted success as learners of maths. It concentrates on four students whose perspectives during the year will be focused on later (Sheila, Michael, Diane, Linda). I shall outline the others' experiences. All names of individuals and schools are pseudonyms.

SCHOOL

Students had a variety of memories of school maths. All shared, however, a near total lack of any experience, as pupils, of the investigational approach. Some had found maths easy but boring, apart from the odd year with a particular teacher, until 'A' level (Emma, Jenny, Martin, Richard, Sheila). Sheila painted a particularly negative picture of her grammar school experience. She had been a "timid" child and had had some "really nasty teachers" who had "picked on her". In her first year, she "was just scared all the time really". Others had enjoyed the subject throughout the secondary school (Linda, Sandra, Neil, Paul, Michael). Linda had enjoyed both her 'O' and 'A' level work, maths being one of her favourite subjects.

Michael, having failed the 11+, seems to have bolstered his self-image via achievements in maths. In the first interview, asked to describe his *experience* of maths at school, he recounted how, as a second year infant, he'd come second in an arithmetic test in spite of running out of ink and not finishing all the questions. This "shook" some other pupils and pleased him. And maths at the secondary modern "was good, sometimes remarkably good: one exam in the third year I got 97%, which was unheard of". He spoke similarly about his 'A' level experience in a technical college.

John also failed the 11+, but regained his confidence through finding maths easy at his secondary modern school where he'd enjoyed the subject because he was "good at it", not because he "loved" it *per se*. From being good at it he gained "respect". "Laziness" led to problems with 'A' levels.

Diane had a "horror" of mental arithmetic at junior school, because of one teacher's practices. She had been "very frightened" of her maths teacher in the first two years at grammar school. Things improved when she found herself with a new teacher, in the second set, in the third year. She was "a nicer person", "a lot more sympathetic: I think I could have coped with the higher level if I hadn't been so terrified, but I got the bottom grade pass at 'O' level". She failed maths 'A' level because, she says, the teacher was absent for six months. However, she enjoyed maths at school. She always felt that given the time and the "right teaching" she would enjoy it at higher levels.

Joan had found the subject difficult at junior school: "there never seemed to be anyone around to help me." She "failed" it in three separate Common Entrance papers, finding fractions particularly difficult. Things changed dramatically in her subsequent girls' private school. Here she experienced an algorithmic teaching approach: examples on the board were carefully explained and copied down, followed by exercises and lots of homework. She no longer had problems with fractions. She studied Additional Maths from the fourth year, finding it a "struggle" but "getting through it somehow." She moved to a GPDST school to study maths, further maths and two science 'A' levels, thinking she might become a maths teacher. She didn't enjoy these. She wasn't a "true academic", only working so that "one day she could stop". It was "hard work", and not easy. Occasionally she thought of leaving.

UNIVERSITY

Those who studied for a maths degree did so in universities. The shock experienced by many students moving from 'A' level to university maths has been documented by Bibby (1985). In his study, students whose typical sixth form experience was of algorithmic maths, of learning how to solve fairly standardised problems for the purpose of passing 'A' level (HMI, 1982), later found great difficulty in coping with the structural emphasis of university Algebra, and particularly with Analysis courses stressing rigorous proof. Furthermore, sixth form work had not generally led them to consider the historical development of maths. They therefore had no way initially of appreciating why mathematicians might have come to believe that it was important to prove, for example, the existence of Real Numbers. They had no way of contextualising the work. The PGCE students with lower seconds told similar stories. Linda said:

> I enjoyed my university life but the maths - just too hard. [BC:Right from the word go?] Yea. It was just different. We covered my A level course in the first lecture and went past it, and being faced with Analysis in the first lecture, in the first term, when you've never seen anything like it before, you just don't understand it and lose interest and then it just goes above your head. Only a few people actually understood most of the maths, and those were the ones who came out with Firsts.

She preferred the algorithmic approach she had found relatively easy and enjoyed at school:

> I did a maths degree. I didn't change to maths and stats, but I did more stats options in the third year. But I preferred the maths really, not the Analysis. But I did ordinary differential equations where you've got a

> formula and you've got numbers and you put them in
> and you get an answer. That's the sort of maths I enjoy.
> I enjoyed the statistics and that again is all the same,
> dealing with numbers, putting them into things and
> seeing what you got out.

Sandra and Jenny gave almost identical accounts. Emma found the first year "nigh on impossible" and the beginning of the next "pretty tough going". She preferred pure maths to applied, finding it difficult but interesting. Things got easier in her third year when she could concentrate on options where visualization was possible.

Sheila, who had taken her degree in 1971, had found it hard:

> Partly the maths was so different. So much of it was
> very abstract and you needed to be able to pick things
> out of thin air it seemed, to be able to produce answers
> - which I couldn't do. Also it was so high-powered.
> There was so much pressure that the quantity of what
> we did and how complex it was made it very difficult
> trying to keep up. There was no time to understand
> lectures while they were actually going on. You just
> had to try and get all the information down and then
> try and figure it out yourself afterwards, which made
> that side of it very difficult. We had some very bad
> tutors, one in particular. You used to get worksheets to
> do and if you couldn't do them you discussed it at the
> tutorials. But some of the tutors wouldn't discuss how
> to do things anyway. It was so obvious that we obvi-
> ously hadn't tried it and he wasn't going to tell us how
> to do it.

In her third year she chose as many non-mathematics options as possible, no longer wanting to be a mathematician.

John had enjoyed his (Pass) degree though finding it "incredibly difficult" and having to "struggle very hard for what I got". He wrote about going to college:

> The euphoria was quickly followed by the shock of
> realising: (i) how difficult the work was, (ii) how much
> work there was to do, (iii) how bright most of the
> students were. I never really recovered from that and
> failed my first year exams that June.

After a year off he resat his exams, passed, and proceeded to a Pass degree.

Hayter (1989, p. 256), a PGCE tutor, has noted that teachers helping to interview PGCE applicants are often disappointed that:

> so few are able to enthuse about their mathematical
> studies at university. Their application forms express
> enthusiasm and interest in mathematics but discus-

> sion usually reveals this to be linked to their success
> and their experience at *school* level.

This would apply to many of the cohort studied here.

Neither has the experience of the three students with firsts in maths been straightforward. Joan "hated" university, finding work "very hard". Paul, the "class swot" at school, eventually found university maths "unstimulating". He underwent a "Pauline transformation", deciding "exams aren't all that important in the great scheme of things" and writing a maths essay based on *Through The Looking Glass* (Lewis Carroll) which "only" got a lower second: "but I had fun writing it and that's what mattered." He took courses on Problem-Solving and Russian, increasingly regretting his "narrow education."

Richard underwent a profound shift in his attitude:

> At university, I loved maths, I wanted to do research.
> Not so much research but to become a university
> lecturer and teach people about maths. We edited our
> own maths magazine when I was an undergraduate
> with the aim of spreading popular, fun maths around.
> But I went to [prestigious university] to do research
> and that was probably a mistake because there was no
> corporate mathematical body to speak of there. I spent
> all day on my own trying to do some work. For me
> anyway that was a failure. No one did maths at
> [prestigious university]. It was all done elsewhere and
> then just disseminated there. So I just got a job. Quite
> a good job I suppose, business analyst at [well-known
> firm].

From solving problems collectively in the pub he had moved to a research environment where the stress was, he said, on isolated individual production. As a result of his inability to cope with this he temporarily lost interest in maths.

Four students remain to be discussed (Diane, Martin, Michael, Neil). None took first degrees in maths, though Diane later took an Open University degree, after part-time study of pre-university maths.

Neither Neil nor Michael had to confront the sort of abstract maths that led to so many of these students' difficulties. Michael had tried unsuccessfully to enter a maths degree as a mature student. Of the maths in his recent Computer Studies degree he commented:

> I'd like to have done far more and to have dropped
> some subjects but in the first term we did an A level
> revision course in thirteen weeks which brought back
> memories. We did virtually an A level statistics course
> in the next term, very very hard work, and I got a grade
> A, first year and the second year, for all my maths. And
> I enjoyed it.

As in comments about school his grades are stressed.

Martin had considered studying maths but chose engineering because of the job prospects (in 1960) and because his peers were doing it. He then did research on electrical machines, finding much of the required maths "quite stimulating". He has maintained an interest in number theory.

Diane enjoyed her Open University pure maths but struggled with the applied. She'd passed, but "not well". She did not, however, enjoy group work during Summer Schools.

Examining these students' previous experience of maths, a pattern of *warming up* at school followed by *cooling out* at university (Clark, 1960) can be seen to be quite common. It is therefore likely that, on joining the PGCE course, some will have felt most secure with the content and pedagogical approach they experienced at school, notwithstanding that several described it as "boring".

Now, obviously, the sorts of mathematical experience many of these students disliked at university and the sorts of investigational approaches now being practised in some schools are qualitatively different. However, they share one major feature: they are both different in various ways from what these students succeeded with at school (in some cases, after overcoming difficulties). Although several said they found the algorithmic maths of their schooling boring, this content and its associated transmission model of learning (Barnes, 1976) had at least "worked" for them. We might therefore expect that some of their initial reactions to investigational approaches will have been partly conditioned by a lack of confidence in the face of a new uncertainty about their ability to cope. Another possibility, of course, is that students have preferred views of maths and pedagogy for other reasons to do with, for example, their value positions about education in general. These possibilities will be explored later. It will be seen that no simple pattern emerges.

REACTIONS TO NON-ALGORITHMIC APPROACHES IN SCHOOLS

The School and Student Case Studies

I shall focus on four students, selected for several reasons. First, three (Sheila, Michael, Diane) are mature students, thus allowing an examination of how an important category of entrant copes with discontinuity between their past pedagogical experiences and investigative work. Second, two (Sheila, Linda) had very negative experiences of university maths. Third, two were initially placed in a school (Exford) whose approach to maths teaching was via investigations. One of these soon left the school for another (Plainside) which kept investigations separate from other work. The other two students were in a school (Hartsfield) with much investigational work but whose maths teachers embraced a variety of pedagogical approaches. One undertook further practice in another school (Wilston) characterized by a variety of pedagogic approaches. It is therefore possible to examine, via these four, students' reactions to schools with different pedagogic emphases. The four and their schools were:

Exford (maths via investigations): Linda + Diane (briefly)

Plainside (investigations boundaried off): Diane (after autumn half-term)
Hartsfield (a variety of pedagogic approaches): Sheila + Michael

Wilston (a variety of pedagogic approaches): Michael (later)

A longitudinal study makes it possible to consider the importance of possible influences on students' expressed attitudes by examining how their views change as a result of experience in schools. Do some students, whose initial reactions seem to be the result of a lack of confidence, come to be more positive about newer approaches if they find they can actually cope with them in the classroom? Do other students begin and remain largely against newer approaches on value grounds? Of course, there are other factors to consider in examining later interview data. In particular, students will have passed through the "honeymoon period" and will be exercised by problems of classroom control and "the search for material" (Lacey, 1977). Investigations will therefore have to be reconsidered by most students in terms both of their power to motivate and their potential for classroom chaos. I shall now examine the reactions of the four students.

Hartsfield: Sheila

Before entering Hartsfield, Sheila, who had been "scared" in grammar school and had come across investigations "a little" in pre-course observation, thought children tended to "find it more interesting" but worried that, "in some ways, it's more threatening for them because they've got to try and think things through for themselves". And, perhaps reflecting her experience, she thought work should be organised in groups to make it possible for quieter children to speak out. But, overall, for Sheila, who had found her 'O' level maths easy but boring, and wanted pupils to enjoy maths because "it's very sad that so many people say they hate maths", the new approaches were a "good thing". She also wanted the study of maths to be linked to "real life". She had just been taught maths:

> to be able to do the exercises. It stood in complete
> isolation from anything. Whereas now it's taught as
> being related to what happens in the world around
> you and ways you can use it in practice. To do with
> things that you might come across in your future life.
> Decorating your house, how much paint you'll need.
> It's a much more practical approach which I think's
> much better.

In her first term at school she worried about having to become more of an "enforcer" than she had hoped, to provide children with a suitable learning environment. Also, responding to children's demands was stressful. Discussing investigational work in December, she said:

> they seem to need a fair bit of steering. They tend to
> sort of stumble halfway through which with different
> things going on in the classroom [makes] it difficult
> getting round everybody because there are so many

> with questions. So that you find there are some who've
> missed bits out or can't find a way or who've got the
> wrong answer. So I shall have have to find some way
> of getting around.

"Unlike some", she didn't find it difficult to be less directive than her past teachers, but was concerned about the rationale for the new approaches:

> I don't find it difficult not telling them what to do but
> if having tried it they're all at sea with what they're
> trying to do then they need some assistance. It's not my
> role in the investigational work that I find a problem.
> It's knowing what they should get out of it.

But then she hadn't, she said, known anything about investigational work, never having done any. She appreciated it was supposed to be more interesting but found some children threatened by it. These would prefer traditional work where "at least they know what they're trying to do". And, while this approach was supposed to give children a better understanding, pupils often failed to think carefully enough about their results, failing to make links between them. But:

> A lot of mathematicians had a lot of trouble discover-
> ing lots of things. To expect children to be able to find
> them out so easily seems unrealistic. But I think they
> have advantages, both sides. The best thing probably
> is to try and get a balance between them, so that the
> ones who like the investigational maths have got some
> of that and the ones who feel safer with the traditional
> maths have got some of that.

Clearly she was worried about whether children were learning what she saw as important mathematical content via investigations, and about children's capacity to make discoveries.

She was aware that her previous experience was making it difficult for her to accept some aspects of newer definitions of school maths. An investigation carried out in a Personal Tutor session had involved joining dots in various grids subject to certain constraints and searching for pattern in the results. Discussing this, she said: "That was just a bit of light relief. To me it's not mathematical, although I know we're teaching maths, and it's supposed to have value for mathematical thinking." This was "just because the maths I learnt was so different." However, she was reflective about the issues involved: "... the investigations are all so different that I don't know that you can really generalize between them. I haven't had any very open-ended ones yet so...".And later, talking about finding relationships and formulae in investigations, she said she now understood, from her Teacher Tutor, that "you can do traditional maths by investigative techniques as well, rather than actually having to give [it] up." She knew things like that now, "which [she] wouldn't have done before". As for the value of an investigational approach, there was much she wasn't convinced about. She wanted to see what children gained in practice, though that was difficult to assess.

Sheila's uncertainty is not surprising given her commitment to making maths more enjoyable and more applicable for pupils, coupled with her own traditional experience at school. She still worried whether an investigational approach suited all pupils.

In February she was still reflective. She wondered why two classes, normally taught by the same "innovative "teacher, had responded so differently to a problem-solving exercise. Also, she'd been told by her Teacher Tutor that she had an investigative style of teaching. This had made her even more "confused" about what "an investigation is". Did it have to be open-ended, for example? She still felt many children found this approach more difficult than a traditional one, but added that "people say it is better for them because it teaches them to think rather than follow rules - improves their mathematical thinking". She was "very undecided about them".

She still felt the university investigation described earlier was not really mathematical. Her attitude to such work hadn't "changed much, really", because: "I was taught maths as being something so different. It's more a way of thinking than what I think of as maths I suppose. But then my definition of maths is probably a lot narrower than other people's." She looked for "content and applications" but saw this investigation as lacking both.

She had worried in September about "finding I've forgotten my maths, because it was so long ago." There were also "new bits" in the syllabus that she "hadn't done at all", which she'd have to pick up on the way. This concern resurfaced when she explained why she'd like to find a post in a school using SMP 11-16. Lacking experience, she found it difficult to continually find her own material. This caused her a lot of worry and could result in "quite bad lessons" if she got it wrong. SMP was tried and tested, and would be a "good basis to work from". It was "so long since [she'd] done her maths" and "it seems to have changed a lot". She added, laughing, that she knew many people didn't like SMP, because "it still smacks too much of the traditional". After teaching a while she might come to agree. Clearly, a problem of confidence existed here, compounded by the pressure of the search for material (Lacey, 1977).

Interviewed after an investigational lesson in March, she was again reflective about her work. She pointed to the different sorts of rules children might derive. She employed a multi-dimensional view of mathematical ability, applying it confidently to named individuals.

In May, she suggested that there was a Sussex bias towards investigational approaches. She linked this with worries about her mastery of mathematical material, wishing there had been more university work allowing her to "catch up": the stress on investigations had excluded everything else, even though traditional teaching continued in schools.

But she was not against investigational work. Asked if any books had influenced her attitude to maths teaching during the year, she said: "I like Mottershead's books. They're ideas for investigations. They're lovely. They're fascinating. Such lovely ideas. They're the sort of things that you feel people ought to get the chance to see and maybe would get a different idea about what maths can be." I noted this book

(Mottershead, 1988) had supplied the investigation she had said wasn't really maths. Her reply was: "I know. I still have sort of a split personality on that. I mean, there are more mathematical things in the books than that. Yes I suppose it's sort of spatial. It's a spatial problem in a sense anyway."

Another book, Orton's *Learning Maths* (Orton, 1987), she liked because it concerned itself with objectives, aims and planning. She was now "more worried" about her grasp of content and ways of teaching it. A textbook-based scheme would help her:

> To start with the major problem was coping with the class. But as I spent more time teaching I became more aware of how much you can affect that yourself by trying to teach them the wrong things or the wrong level. That in a sense content is even more important. Not just from what you're trying to teach them but also from the way they're going to react to what you're using to try and teach them. So that made it more difficult. 'Cos a lot of pupils that I spoke to at my son's school who were using SMP liked it. And Richard has been at Wilston and he says they like it. So, although a lot of teachers have a lot of comments against SMP, as long as you introduce enough other material and don't rely on it as meaning that you don't have to do anything else, then it's probably quite good.

In discussing investigations, she referred to Linda's comments about Exford:

> It's aimed at teaching you, helping you how to solve problems, isn't it? So I read. Yes, with investigations in with everything else it is good because it does help to get you to try and bring different things together to use, and solve things. And it gives variety in what's being done. But to use it virtually to the exclusion of anything else is, must be, could even be a damaging approach. I mean the way I've heard it's being used at Exford. Seems to me to be totally negative to them learning anything.

It seemed the children weren't even told "if they're totally wrong" so "how can they be expected to learn anything right"?

In Sheila's case we seem to have someone who, in spite of her algorithmic background, and perhaps because of an overriding commitment to make maths interesting and enjoyable, has been able to keep a fairly open mind about the relative merits of different approaches. She has perhaps moved towards favouring the new, though not at the expense of the disappearance of more traditional teaching. Confidence about her mastery of content seems to have been a major concern. This led her to deprecate what she saw as an over-emphasis on investigational work in the university. She also was worried about the value to pupils of what she saw as less applicable investigations.

Hartsfield + Wilston: Michael
Michael's views on investigations - about which, initially, he "had no idea at all" what people do - soon developed negatively.

In September, after two weeks of observation and teaching in a pedagogically traditional school but before entering Hartsfield, he described his surprise at some children's failure to "remove brackets". This was "the sort of thing I remember being taught and taking straight in". The children were "keen to learn but incredibly slow - not being able to know what an even or odd number is at fifteen - it's frightening". A concern with "basics" appears, but, perhaps because he knew little about investigations, he seemed open-minded:

> It's horses for courses. If by moving away from chalk
> and talk you're going to make the mathematical lit-
> eracy of this country better, anything to do it, because
> if one believes what one reads in the paper, it's going
> downhill. I enjoyed maths, "chalk and talk". But it
> didn't suit everybody.

In December he described his practice school thus:

> So different from when you and I were at school.
> There's far less chalk and talk and more doing, and this
> business of flexibility. I'm not used to it. I've got to
> adapt. What with [coursework] and lack of discipline
> as I knew it, it took me half a term to get used to this
> completely different situation. I'm still finding it hard.

It was also hard to accept that some children not only "don't like maths" but "can't see they ought to do it" because "it's going to be needed for life".

It appears that both confidence in and a preference for algorithmic approaches were influencing his attitudes:

> Disappointment's too strong a word [but] for four
> weeks, with two classes, all they've been doing is
> project work. Nothing exciting, not how I remember
> the maths. Not, as the second years do, gradually
> working through exercises, seeing that they're doing it
> all right, going round. More, sort of, such open-ended
> things that I don't know.

> - and I've got to think because it's GCSE, no I can't, I
> can guide but I can't help, or can I or should I or what?
> I don't want to actually give the words but I want to
> give the flavour. It's experience. Whereas, to do a chalk
> and talk thing, it's something that I've seen done at my

school years ago and at [observation school].

But he was enjoying investigations in university Curriculum Tutor groups. They were a "challenge, possibly because I haven't done maths as a degree". One, involving the making of curves outdoors with people as points, had produced "amazing results". He had followed it up, using Calculus, "his favourite". His problem was the classroom implementation of such approaches:

> It's the part of me that isn't very inventive. Not that I'm lazy. If somebody says to me, do something, I can do it. But if they say use a bit of imagination, like a day in the life of a penny: write an essay, I'm stuck. But write a review of a book, or something where you've got some boundary, I'm far better. In other words chalk and talk to me is easier.

He was uncertain about his classroom role. In university Curriculum Tutor "lessons":

> The other half is an investigation by us, which I find most enjoyable. No, I'm not going to contradict my-self. I find it enjoyable because I'm doing it, but when I'm at school there's nothing to mark and therefore I feel almost redundant for fifty minutes. Sometimes, they just sit there, and if they don't want me they just work and I feel redundant as a teacher because I don't want to provoke, I just want to check every now and again that they're doing the right - and if they are, I sit down. I feel it's not right. So that's how I feel, doing coursework, er, investigations.

In February, asked to describe a successful lesson, he chose one involving a structured piece of discovery plus "chalk and talk" and textbook exercises. They "seemed to understand it, I enjoyed it and they did all the exercises in one of the books". With this sort of lesson, he could check examples in advance, "actually work them all out so that I know what I'm doing", and write notes ready for the blackboard. This "help(ed) my confidence".

Describing less successful lessons, he first said these resulted from lack of preparation, but then concentrated on other causes. The outdoors geometrical investigation he'd enjoyed at the university had been a "complete and utter disaster", with the children "not paying attention" because, he thought, they didn't see the work as maths but as "messing about". But they were an E set, "not graduates", and he'd never do this again with "those classes", adding that most of his problems were "caused through discipline", about which he was very con-cerned. Here he displaces blame (Lacey, 1977) for difficulties downwards to the perceived nature of the pupils. This displacement, and concern about his role in coursework, featured in many remarks.

Comments about marking suggest that the uncertainty involved in less algorithmic teaching was a major concern. He'd had to assess a piece of work which involved deciding whether various objects and materials would be better measured by weighing or counting. He was concerned by the "vagueness" of some of the questions - prepared by another teacher and which, "regrettably", he'd handed out without looking at them:

> The idea was if you had butter it would obviously be
> weighed. Cards in a pack would be counted. And then
> peas, are they counted - peas? Or do you measure
> them? And it was a little bit vague. And one said how
> do you count the length of a match? Most people put
> down millimetres. But one chap wrote minutes. Obvi-
> ously thinking of football or cricket. Bit vague. That
> was the problem.

His experience of maths doesn't seem to have prepared him for this.

He enjoyed work which allowed him to proceed in ways he understood and didn't involve disciplinary problems. He especially liked "the one to one relation-ship, well, the one to few relationship", for example, work with a small group of sixth formers. It was the only time he'd "ever heard a pupil say 'Thank you, sir', rather than ugh". Normally, he spent more time worrying about discipline than maths. This was "very frustrating". Children in bottom sets were "so variable, so volatile". He was frightened of "losing his rag", having done so once or twice. Schools had changed, and:

> possibly again, it's a comprehensive in [town]. I don't
> want to teach in a comprehensive in [town]. I'm
> applying for jobs in grammars. Although I'm totally
> opposed to grammar schools from a political point of
> view, that is one area that shouldn't be quite so bad -
> the not wanting to learn maths.

Displacing blame downwards, he describes a way of avoiding problems via choice of school.

He was still finding it difficult to make sense of investigations. He didn't see what "bottom sets" were "supposed to be getting out of it". It seemed to be "just something to give them to keep them quiet". (Others might have interpreted this as a positive motivational strategy.) He could see what one recent project had been about, but not another on loci. He didn't really see how "loci work", which he'd not done at school, "really comes into maths, apart from ellipses and circles, that sort of thing".

He was still very "disappointed" about "basics", like "tables". He found it very difficult, trying to explain, while "you've got 23 others." He wanted "to get to a school where the kids will appreciate me teaching maths". Many problems would then disappear.

In March I observed a very algorithmic "example + exercises" lesson. Pupils were told that "cosine equals adjacent over hypotenuse", one example was worked on the board, and they were set individually to do similar examples from a worksheet (all triangles being similarly aligned). His remarks about the worksheet did not suggest that much pedagogical reflection characterized his teaching. It seemed more a matter of getting by: "To be honest, if [Head of Dept] gives them to me I hardly look at them. I've used them before. It's very reliable. I didn't realise that question thirty had a mistake in it. I wish I'd known that." Asked whether one run-through the example and a partial rerun were enough for pupil understanding, his reply suggested an algorithmic perspective:"...they were able, as far as I was aware, to do it, therefore I had explained it well. But whether it actually sunk in, I don't know." Nor had he discussed sine or cosine in relation to the ideas of enlargements or similar triangles - ideas used in traditional "good practice" to help make sense of trigonometrical ratios.

The school put him "at risk" four days later. When interviewed in May, his main concern was to pass the course (he was to undertake further teaching in a school generally seen as easier discipline-wise). He was also looking forward eventually to working in the girls' grammar school that had offered him a job at about the time he was put "at risk". His comments on investigations reflected a belief that he would eventually have to cope with "coursework", because of the GCSE. He was "happier doing traditional maths", but felt that if he could go into a school that was not doing coursework for another two years, and change with them, with training, then he would cope. At Hartsfield he had had to cope with both "the school" and "this new method of teaching". He still "couldn't relate to this" method, though he was "better than last time". He lacked confidence in his ability to run coursework lessons, not knowing how to motivate the children "for such a long period", how to "set goals for three weeks". In particular, he felt "very insecure teaching project work" in mixed ability classes.

Other remarks suggest that he was also somewhat unreflectively wedded to the definition of maths that he'd experienced at school. This was demonstrated by his comments on multiplication tables. Children should learn tables up to *twelve twelves*, as well as long multiplication and division, "because that's the way I learnt so that's the way I presume they learn tables today. By rote". He still displaced blame for problems downwards and remained optimistic that many would disappear in a grammar school. He had "only spent four days in grammar schools in his life" but had seen "kids sitting there almost saying please teach us maths". Furthermore, there would be a higher level of motivation and parental support, whereas, at Hartsfield, particularly with the bottom set, "they couldn't care a shit and probably the parents don't either".

I next interviewed him, his future still uncertain, in June. His second practice school was:

> A different world. The kids. I was able to spend more time thinking about the maths and the structure of the lesson rather than have to think [about] discipline problems and trying to get their attention all the time. That's all classes. I was able to take lessons from Hartsfield which I'd hoped would work well there, and transfer them. They worked very well.

He "was amazed" how well the outdoor geometrical investigation, which had been a "disaster" at Hartsfield, had gone:

> I admit there were only twelve of them. But there was
> no natter. They paid attention. They were keen. They
> were enthusiastic. And that was lovely.

He said his confidence had grown at Wilston. He hadn't been seen, he thought, as a student teacher. As a result of one lesson "going well", the next would be easier, and so on. They knew "his standards". (At this point he'd done very little investigational work in Wilston.) His confidence had been helped particularly by his having been able to "lecture" to fourth years on a familiar topic:

> ...another thing that's helped boost my confidence was
> Dave Jones asked me to teach some IT lessons. This is
> a subject more recent to me, and I was able to really feel
> at home. And they enjoyed it because I was able to tell
> them examples. The first one I did was input and
> output. I got a lot of lesson notes from my lecture notes
> of three years ago. I had them all prepared, and I was
> able to show them bar codes, charts - they thoroughly
> enjoyed it. And I taught in a more lecture way where
> I had some lecture notes typed out. I had re-edited
> those to short notes, and put those on a glossy. And
> they copied those down. We went through a couple of
> input devices. Where I gave them the spiel, examples
> - bar coding, the advantages of it. This was an area I
> was far more at home in. I asked them questions like
> what is this? What's it for? Then they wrote up my
> notes. The next week they sort of were smiling, "what
> are we going to do this week?" Rather than "oh, we're
> not doing that again."

He was increasingly sure that his problems had resulted primarily from Hartsfield, though "I made mistakes". He hadn't liked the "feel" of the school. They didn't have a school uniform. You had to "fight to get attention". It "wasn't how I remembered school". Not all teachers were "made" to teach in such schools. The head of his grammar school-to-be agreed.

We can see that someone who was initially unconfident about his ability to cope with what was, for him, a dramatically different way of teaching remained concerned at the end of the year. He believed he would be able to cope with coursework and investigations eventually if proper training was provided, if in a school that was changing with him, and if the pupils were better motivated than those he'd experienced at Hartsfield. But he had not come to understand the rationale for such approaches, and was still unclear about his role in investigational lessons. He remained happier with algorithmic approaches. He had, however, enjoyed doing investigations at the university.

Michael's confidence has taken a severe blow this year and, given this and his previous investment in algorithmic school maths, plus his wish to have a leading role in the pedagogic process, it doesn't seem likely that he will quickly embrace newer approaches.

Exford: Linda

When first interviewed, Linda had been at Exford for two weeks. Maths in Years One and Two seems to have been more or less entirely based on investigational work. For various reasons Linda and Diane were soon offered places in another school. Diane accepted, Linda did not.

Linda had expressed a preference for more algorithmic approaches when discussing her university experience. However, in late October, she argued, of investigational approaches:

> It's good. I like it. It's difficult to say that cos I've been at Exford but it's quite good to see a bit more of that and not just textbooks [Why?] Well it gives variety more than anything. But they do it too much. [Why too much?] They do all investigations. [Why is that too much?] The kids are bored with it because that's all they do. And also they've got all these textbooks [and] they don't use them. They've sort of thrown them out of the window. There should be a compromise: you do the investigation and then back it up with a bit of textbook work [because] some kids don't get anything out of investigations. [Why not?] You give them an investigation and they sit there and then they look at it and then they copy their neighbour's and the answer goes round the classroom and they write it down. They don't know why it's the answer. And they get lost. They tend to use big numbers, complicated draw-ings, and get bogged down in the routines and not actually learn the basics. [The basics being what?] Well, it depends on the investigation. ... Some of the teachers actually give them structured investigations but the general trend and the way they like to do it as a department is give a question and let the kids find out what to do.

In recounting what led to the offer of a school transfer, Linda's remarks suggest that worry about mastery of the mathematical content embedded in investigations was a key factor influencing her views. After claiming that attempts to find out more about the school's pedagogical practices were seen as criticism, she continued:

> We sort of turned up [in classrooms] for an investiga-tion and didn't have a clue what was going on. We weren't told and it makes you feel really threatened when you don't really understand. The kids ask you questions and you can't answer them cos you don't know what's going on. All the teachers ever said was

> ask the children what they're doing [but] children
> don't explain things properly and in the investigations
> the kids don't set out their work nicely so that someone
> can come along and understand it, and it got very
> confusing. It's OK if you're doing fractions on the
> board, you can remember that, but when you're doing
> these investigations you really don't know what's
> going on. The teacher can be very vulnerable.

Alongside vulnerability there was another factor. Maths teaching should involve
the transmission of definite content:

> Probably not necessarily a definite right answer but
> there's got to be some meaning the kids get out of
> investigations. There's got to be something solid they've
> learned. And if they're completely off the track then
> the teacher should be able to lead them. If you set an
> investigation you've got to have some objective in
> your mind that you think they're going to get out of it.

She would be less concerned if investigations were "one-offs". But, at Exford,
children were "learning their maths by investigations". It was "all very well, that's
good", their going off in different directions in an investigation but "there must be
something in there [so] that they're actually learning some maths".

In February, she described a successful investigation:

> First years, using polydrons. The little plastic shape
> things that you fit together. That was team teaching. I
> taught with their teacher. But the activity went really
> well. They loved it. It was also something that the
> slower ones could do easily, just as well as the brighter
> ones. If not better cos sometimes they're better at that
> sort of thing than the brighter ones that can do all the
> sums and that.

It went well because "it was something different", with the children "fiddling and
making things themselves", "rather than writing all the time". Linda was continuing to see pedagogy in terms of motivational consequences. She moved to a negative
example, a less successful middle set fourth year lesson. The material, combined
transformations, was "too hard", so they had "no motivation". She hadn't known
how to "break it down to make it easier". They were "bigger than me: I'm scared
of them - I don't know if they sense that".

Clearly, as well as worrying about teaching investigationally, Linda was worried
about discipline (Lacey, 1977). She was also "still threatened" by fears relating to her
role in the investigational classroom. She wondered whether she shouldn't be
directing pupils more towards discovering standard mathematical content, but had
begun to find ways of coping:

I'm learning to waffle out of it. But I get very paranoid
when there's several things going on. They're all doing
different things and you think, where am I going?
Should I be directing them? And when some of them
are doing things that aren't really going to lead them
anywhere, or aren't really mathematical, whether I
want to bring them back to go my way.

But, though coping, she was finding it difficult to accept some definitions of
successful mathematical activity. One lesson, involving making triangles from
pieces of string, had not gone well. The children "weren't sticking to a pattern, they
weren't getting anything out of it". But the teacher thought "it wonderful they'd
done all these different things and all these ideas on their own". Linda couldn't
understand this view.

But, by February, she had accepted a definite, if limited, place for investigations.
They should be backed up with textbook work. They should be used more rarely,
"one a term or something", so that pupils might "put more effort" into them: they'd
"see it as something different rather than something the same again and again". She
turns her concern with motivation against reliance on investigations.

She was particularly worried about investigations in mixed ability classes. The
"lower ones" needed more time than "brighter ones" but weren't getting it because
you "have to stop the whole class together". "Brighter ones" were always "clamour-
ing for you to say what can we do next". Often "lower ones" just copied from the
others. Setting was emerging as a way of alleviating anxiety about what children
gained from investigations. The third year were set, lower sets being given "easier
investigations, and more time, and they seem to benefit more".

In March, Linda saw more "good points" in the investigational approach
"because I develop as I'm getting through the year". But Exford went too far:

I'm teaching enlargements. By investigations, and I
don't think that that's necessary. The workscheme, it's
a line, says draw two triangles on the board, one an
enlargement of the other, ask questions. I don't know
what kind of questions. I mean, what you had for tea
last night or did you clean your teeth this morning?

Worry about how to carry out the teaching task and her pre-course preference
for algorithmic approaches are perhaps both operating here. But, by the end of May,
Linda seemed to have decided that investigations were useful. She had withdrawn
a job application because the school used SMP booklets:

I don't like the thought of booklets. All the time. Just
being an organizer rather than a teacher. I don't like
the idea of a booklet saying it all and you just give it out
and then collect it in and just tick them. I know not all
SMP's like that but for the first two years it is. And
that's the years I enjoy teaching most. You get so much
input, especially some investigations. You get so much

> input from first and second years. It's by the time it
> gets to the third and fourth years that I find it's a bit
> stale.

The issue of motivation seems to have been a key influence. She'd also found various publications on investigational work in schools helpful and reassuring during the year. These described:"...what investigations were for. They've also got examples of children's work, which helps an awful lot. Seeing that pupils from other schools are doing just the same sort of work as the kids you're teaching. It's really helpful." But she still disapproved of Exford's pedagogical emphasis. If she taught next year: "the main thing would be to learn how to teach. Rather than dish out investigations all the time. Learn how to use textbooks and how to produce worksheets and that sort of thing. Which I've had no experience of." She preferred investigations that "led somewhere". She preferred the term "problem-solving". She repeated the story of the string:

> If you've got a really good problem to solve, it's really
> good. But some of the things I was actually doing,
> what they called investigations, were sort of a bit
> wishy-washy. I didn't think it was achieving anything.
> I did a thing about string. Which in its own [sic] is quite
> good. But there's nothing necessarily to lead off from
> it. When I did it all the pupils led off in different
> directions and I didn't think they learnt anything
> about maths in a few lessons. But the teacher said it
> was wonderful that they all did different things. That
> sort of thing. I just don't see where it's going. That was
> the one that really got to me.

Her perspective on investigations was still very much coloured by her initial algorithmic preference. Investigations were seen as new means, sometimes motivationally effective, to the traditional end.

Exford + Plainside: Diane

Diane, the Open University graduate, shared many of Linda's concerns. Before entering Exford, she was apprehensive, having heard the school was "progressive":

> I think they mean they do a lot of investigative work
> and they don't use textbooks very much. That fills me
> with alarm because that's not my experience of maths.
> I've met problem-solving and investigations through
> my children but not really first hand except during the
> OU summer schools when we do group work. I hate
> that [because] I find it difficult to think at the same
> speed other people are talking. I tend to feel panic that
> I'm the only one who doesn't understand what's going
> on. Yet from experience I know that if I can go into a

> quiet corner by myself I can do it, but I can't cope
> initially in a group setting.

However, her children enjoyed investigations and she saw, from pre-course observation, that such approaches might have motivational advantages:

> I did my observation at [working class school] and the
> children seemed very bored, and had a lot of negative
> feelings towards maths. They hated the maths lessons
> and they hated the maths teachers. I can see that
> making maths more interesting is a move in the right
> direction but I don't know how I am going to cope with
> teaching that way.

A transmission-orientated concern, whether strategic or internalized (Lacey, 1977), also appears in her remark that, as a probationer, she'd "be worried about not getting through the syllabus because we were playing around with investigations." But confidence seemed to be a key factor. She would "feel safer with a textbook" and would be very threatened by having different things going on in a classroom.

In November, comments made when giving her version of why she'd left Exford seem again to point to confidence as a crucial factor in her reactions. A focus on product rather than process is also apparent:

> The problem was the approach to maths teaching.
> Because they teach maths solely through problem-
> solving. They don't use textbooks and they don't use
> banda sheets. They don't teach from the blackboard.
> Sometimes I can see where the lessons are going. I can
> see they're related to area or translations or fractions
> but most of the time they seem to be investigations for
> investigation's sake. This was alright [while] I was an
> observer but I started to get worried when it came
> nearer to the time I'd be teaching. I think it was because
> I was worried. Also if you talked to people in the
> maths department they became rather defensive - if
> you tried to criticize them. Well, I'd say I'm not
> criticizing but it still came across as a criticism. Be-
> cause I didn't understand how their maths teaching
> worked and I wanted to understand. Because I wasn't
> thinking of changing schools. I was thinking of mak-
> ing it work. And I can see some of the virtues of
> problem-solving. A lot of Exford children say they like
> maths.

At her second school, "investigations" were boundaried off, occurring in a special room fortnightly. She was happier:

> The approach is a lot better. You can actually teach a
> lesson on fractions or translations or trigonometry,

> and that is not considered directive or evil or wrong.
> The discipline level's about the same, which is accept-
> able to me. I was a bit worried about this because of my
> previous experience of observation at [working class
> school]. I realised I couldn't teach at a school like that.
> Plainside wasn't like that. So the kids are all right. The
> maths teaching's alright.

She was much less threatened by this school's practices because they foregrounded
algorithmic maths teaching:

> Each class has a double period fortnightly where
> they're in an investigations *room* with all the equip-
> ment. Yesterday I spent the whole day in there. I feel
> quite happy about it because the investigations are in
> context. They're not the only approach to maths. They
> *teach* maths. From textbooks, from the blackboard.
> You can give the children information. It's not wrong.
> And they're using the tools of the maths trade in the
> investigations. Whereas at Exford they were flounder-
> ing. They didn't have the tools. Their ignorance was
> appalling. There was a top fifth year who didn't know
> how to plot a graph. I heard complaints from the
> science department. One girl asked me if I'd explain
> the relationship between pi and the diameter to her. I
> had to do this behind my hand in secrecy because that
> was being directive. That was giving information. I
> wanted to teach the skills I felt I had. Incidentally, I'm
> not any good at problem-solving and puzzles. I'm not
> the sort of person who does puzzles for pleasure. But
> I do see that they're a valuable part of the course.
> Anyway, they have to do them for the GCSE. They
> don't pose any threat to me now, because they are in
> context. They're just part of the maths teaching and not
> the be all and end all.

This school seemed to offer her a way to cope with the new pedagogy. However,
in February, she was again unhappy in investigational situations, feeling insecure:

> I don't like these sessions. I don't feel in control of what
> they're doing. They choose a card from a box labelled
> third year investigations. And I'm not familiar with
> them. I dread them saying what do I do next? Is this
> alright? Because I don't know! And I've always had a
> bit of a fear of investigations because they're quite new
> to me. And Exford was enough to turn anybody off
> investigations for life. Also the third year [are] always
> more rowdy and difficult in the setting of the investi-

gation room. Because they're in groups of four. And they're allowed to get out of their seats and wander around. They tend to get very noisy.

Clearly, Diane was finding it difficult to cope with the uncertainty generated by her lack of knowledge of these investigations. Her belief about the relation between investigations and discipline added to her concern. However, in September, Diane had commented that she could see the importance of making maths more interesting, but worried how she'd cope with new approaches. At Plainside, where "investigations are in context", she does seem to have begun to conquer her fears. Although still worried about the open-endedness of some investigations, she begins to see ways of increasing her confidence, by attacking one area of uncertainty. She even criticizes the separating off of investigational approaches:

> I haven't taken any of the investigations home and worked on them myself. Simply because I don't have time. There's always more pressing matters. That's what one needs to do. And I can't really get involved in any one investigation during the class. I feel I could get interested in some of them. But I'm also worried because there's no right answer. They're open-ended. That does bother me. I like to know where I'm going and where I've achieved it. That's just a personal thing. So I'm never as happy in an investigation lesson as I am in a more traditional one. If I could have my way I'd use investigations to lead into a subject, which is something they don't do at Plainside because investigations and subjects are separate. I think they're trying to drag them in because they have to for GCSE. It's a bit false really cos there's this barrier between the two.

She was now happier working in and with groups, as long as she "was confident in the subject matter: if I knew what I was talking about I was alright". She didn't like "floundering around". In fact, Diane, who had a "horror of mental arithmetic" at junior school, was frightened of her first secondary teacher of maths, failed 'A' level and has only in recent years proved to herself through part time study that she can succeed at maths, seems to be gradually developing confidence in new areas of practice. By May this process had progressed:

> I feel a bit happier now because my daughter, she's fourth year, has started doing a couple of investigations. And I've spent quite a lot of time working on them. I shouldn't say that, should I? I mean, she's doing the work herself but, you know. And I'm really starting to get into them. I think I've just got to sit down and do some. Which I haven't been doing. I think it's just I'm not very familiar with them but once I get into them I think I'll enjoy it. I think I might almost enjoy

that more than straight maths teaching. [Sounds like a
matter of confidence?] Yes. Very much so.

We can see that both the students initially placed in Exford found ways of coping, although in one case after a change of school. For both, though, concerns remained about more open-ended investigations. Linda was worried whether pupils were "learning maths" (a body of accepted content and skills) via such work. Diane was concerned, "a personal thing", about the absence of right answers. This seemed to have much to do with insecurity about the mathematical content of investigations. Both wanted investigational work used in conjunction with more algorithmic approaches.

The three "Firsts" in maths

I briefly discuss these three students to illustrate what well-qualified recent graduates made of the new approaches.

Richard initially thought investigations "great" and generally, as he developed an increasingly critical view of schools, continued to do so. He became very interested in writers like Holt (1973). He explained that, even in the first year at grammar school, where he had "come top in about seven subjects", he had felt great sympathy for pupils "coming bottom". He was against that form of assessment "because some people failed at it". Then he experienced problems as a research student, finding it hard to cope:

> as I began to have problems I began to be against [such
> assessment] because I hadn't had problems. I should
> have done. There's something seriously wrong with a
> school that can let people go through life without ever
> worrying about anything. So that changed the empha-
> sis, added a new dimension to it.

He therefore didn't want pupils assessed against "impersonal scales". He saw an investigational approach as well-suited to matching work to individuals' "abilities" and as making mixed ability lessons, which he favoured, more feasible. He had found the challenges he wanted in rock climbing and similar pursuits. By May, he was passionately advocating more outdoor education. As for maths, he'd like to:

> almost lose it as a subject entirely along with all the
> other things. Just let the children learn and sometimes
> they came across mathematics then there's a maths
> teacher there as a resource for them, to help them. But
> there's something totally alien about cramming 30
> children of the same age into the same classroom and
> talking at them about something most of them aren't
> interested in. They should be out rock-climbing or
> doing something. A lot of the time.

Paul, had mixed, but largely positive, feelings initially. He tried various innovative approaches during the year, including vectors via *The Hobbit* (Tolkien, 1937) and monetary maths via *Neighbours* (a TV soap-opera). He also, in a self-critical way, tried various investigations in his classroom. By May, he saw much good in investigations but worried whether pupils put enough effort into writing reflectively about their work. He'd learnt the importance of this, he said, in his undergraduate "problem-solving" course.

Joan had been horrified by her experience, in pre-course observation, of a school that took a largely investigational approach. She felt the students didn't learn "basics". In her practice school, where some teachers favoured investigations, she continued to worry about "one-off lessons" and the resulting "lack of continuity". More "structure" was needed, a less "wishy-washy approach". Letting pupils learn "through trial and error" wasn't teaching. They might as well "guess at things at home" (February). But she wasn't totally against investigations, "for backing things up". She came nearest, initially, to reproducing the arguments of some of the New Right about school maths. Her views didn't change significantly, but, by June, having experienced a new school as a result of having been put "at risk", she accepted that "they've got to be done" eventually, given the GCSE requirements. Her major concern was finding enough material.

Joan is a clear example of someone strongly wedded to traditional definitions of school maths. This may have much to do with her own first experience of success having occurred, after a period of difficulty, via the algorithmic approach, and with the great *investment* of work she put into such maths. She doesn't seem to have been threatened by pupils' questions in investigational lessons. In May she described such lessons as easier to prepare and supervise. She quite liked the "atmosphere" in them. But in June she again stressed her preference for traditional approaches, saying how good it was that her new school used textbooks. This meant you did not always "have to start back from the beginning; you'd got the guarantee they'd done a certain chapter". In Joan's case opposition to investigations seems to have sprung from basic values about the aims of maths education.

CONCLUSION

There seem to be two main sources of opposition to and concern about investigational approaches: a lack of confidence about content and/or role, and strongly held views about the correct nature of a mathematical education. Diane's initial concern has begun to evaporate later in the year as, helping her daughter, away from the pressures of the classroom, she begins to enjoy investigations. Sheila is perhaps moving in a similar direction, wanting a mixture of approaches, though she seems to have found it quite difficult to adjust to the new definition of mathematical content that, for her, some investigations seemed to involve. She wasn't threatened by a less didactic classroom role.

Linda's attitudes were coloured by her being at Exford, the very pro-investigational school. Initially she felt very threatened by her lack of advance knowledge of the maths implicit in what the pupils were doing. Later, as her confidence grew, she stressed the capacity of investigational work to motivate younger pupils but, reflecting her view that important content needs to be thoroughly covered, she preferred less open-ended investigations and would want "problem-solving" (her preferred term) to be used alongside textbooks and other material. She still needed to "learn how to teach" as opposed to "dish(ing) out investigations all the time".

All these students' responses were coloured by concern over classroom control, especially marked in Michael's case. He preferred an algorithmic pedagogy but also believed that his problems during the year arose because he had had both to teach pupils seen as very difficult and to cope with "this new approach".

Nothing in this paper is meant to imply that well-qualified recent graduates would necessarily be more likely to prefer to work investigationally. That this is far from certain can be seen from the brief accounts of the three "firsts". Joan remained opposed throughout the year.

Given the current shortage of maths teachers, entrants are likely to continue to be drawn - perhaps increasingly - partly from amongst those whose own experience of maths has been problematic. While some of these, like Sheila and Diane, may have sympathy for attempts to make maths more interesting, enjoyable and fruitful for pupils, they may nevertheless, through a lack of confidence, initially tend to want to reproduce the pedagogic strategies they feel most secure with: those they experienced as pupils. They may, with experience, come to see the potential motivational advantages of investigational approaches but, in future, this possibility will be constrained by the National Curriculum targets. Scott-Hodgetts (1988, p4), in interviews with teachers who had been working in investigative ways and had read the report of the Mathematics Working Group, found they believed they would have to take back the responsibility for learning from pupils in order to meet the requirements of the national curriculum.

As far as the PGCE course itself is concerned, the results point to the crucial nature of the students' first experiences of investigational approaches in practice in schools. Clearly, it is very important, and many teachers in the schools recognize this, that students have the chance initially to develop confidence with such approaches through work with individuals and small groups. The university Curriculum Group sessions did allow students to work investigationally from the beginning but, as was seen with Michael, there was no necessary transfer of either confidence or enjoyment from this context to that of the school. The evidence of this research suggests that the university also needs to consider whether, alongside the practice of these approaches, more discussion of the rationale for them might not help many students cope. Such discussion would need to develop students' understanding of the nature and purpose of the various types of investigational and problem-solving approaches used in schools, from the relatively closed, with a "right answer", to the relatively open-ended (Lerman, 1989). It would need, given some students' concerns about "one-off" investigations, to consider how discrete investigations, as opposed to a permeating investigational approach, might fit into an overall set of planned learning experiences for pupils of various "abilities". It would also need, again given some students' worries about this, to consider the different classroom roles a teacher might need to adopt as a consequence of different types of investigation being used with pupils. While such discussion is clearly not a sufficient basis for classroom success with such approaches, it might nevertheless be a necessary one for students whose preferred definition of school maths is initially algorithmic and who therefore lack confidence in the face of investigative work.

ACKNOWLEDGEMENTS

I would like to thank the thirteen students, their Curriculum Tutor, and their practice schools for their co-operation. Carolyn Miller provided especially helpful comments on an earlier draft of this paper. I would also like to thank Neil Bibby, Stephanie Cant, Martyn Hammersley, Colin Lacey, David Pennycuick, Chris Shilling and Trevor Pateman for their comments, and my daughter Kate for renewing my interest in investigational mathematics.

REFERENCES

Baldwin, S   (1926)   *"On England"* Penguin, Harmsworth

Broadfoot, P   (1986)   "Assessment Policy and Inequality, the United Kingdom
                              Experience".
                              British Journal of Sociology of Education 7.2 pp l35-154.

Broadfoot,P.,   (1988)   "What Professional Responsibility Means to
Osborn,M.   Teachers: National Contexts and Classroom
Gilly,M., and   Constraints" in *British Journal of Sociology of Education 8.3*

Abraham, J. &
Bibby, N.   (1988)   "Mathematics and society: ethnomathematics and a public
                              educator curriculum",
                              For the Learning of Mathematics, 8, 2.

Barnes, D.   (1976)   *From Communication to Curriculum,*
                              Harmondsworth: Penguin.

Bibby, N.   (1985)   Curricular Discontinuity: A Study of the Transition in Math-
                              ematics from School to University, *University of Sussex Educa-*
                              tion Area Occasional Paper, No. 12.

Clark, B.R.   (1960)   "The `cooling out' function in higher education",
                              American Journal of Sociology, 569-76.

Cockcroft Report,
Great Britain,
DE S   (1982),   *Mathematics Counts,* London: HMSO.

Coldman, C.
 & Shepherd, K.   (1987)   "Mathematics", in North, J. (ed)
                              The GCSE: An Examination, London: Claridge Press.

Cooper, B.   (1976)   Bernstein's Codes: A Classroom Study,
                              University of Sussex Education Area Occasional Paper, No. 6.

Cooper, B.   (1985a)   "Secondary mathematics since 1950: reconstructing
                             differentiation", in Goodson,
                             I.F. (ed) *Social Histories of the Secondary Curriculum*,
                             Lewes: Falmer.

Cooper, B.   (1985b)   *Renegotiating Secondary School Mathematics*,
                             Lewes: Falmer.

Delamont, S.   (1989)   "The fingernail on the blackboard? A sociological perspective
                             on science education",
                             Studies in Science Education,16,25-46.

Ernest, P.   (1989)   "The philosophy of the national mathematics curriculum",
                             paper presented to Social Perspectives of Mathematics
                             Education Group, June 1989.

Furlong, V.J. et al (1988) *Initial Teacher Training and the Role of the School*,
                             Milton Keynes: Open University.

Great Britain,
DES   (1980),   *Aspects of Secondary Education in England:*
                             Supplementary Information on Mathematics. London: DES.

Great Britain,
DES   (1988),   *Mathematics for ages 5 to 16*,
                             London: DES/WO.

Great Britain,
DES   (1989),   *Mathematics in the National Curriculum*, London: DES/WO.

Hayter, J.   (1989)   "*Becoming a mathematics teacher: grounds for confidence?*" in
                             Ernest, P. (Ed) Mathematics: the State of the Art,
                             Basingstoke: Falmer.

HMI   (1985)   *Mathematics 5 to 16*, London: HMSO.

HMSO   (1987)   *Better Mathematics, the report of the LAMP project*,
                             London: HMSO.

Holt, J.   (1973)   *How Children Fail*, Harmondsworth: Penguin.

Howson, A.G.   (1989)   *Maths Problem: Can More Pupils Reach Higher Standards?*,
                             London: Centre for Policy Studies.

Lacey, C.   (1977)   *The Socialisation of Teachers*,
                             London: Methuen.

Lerman, S.   (1989)   *"Investigations: where to now?"* in Ernest, P. (ed)
Mathematics: the State of the Art, Basingstoke: Falmer.

Mellin-Olsen, S.   (1987)   *The Politics of Mathematics Education*, Dordrecht: Reidel.

Mottershead, L.   (1985)   *Investigations in Mathematics*, Oxford: Blackwell.

National
Curriculum
Council   (1988)   *Consultation Report: Mathematics*, London: NCC.

Noss, R.   (1989)   *"The National Curriculum and mathematics: a case of divide and rule?"*, paper presented to Social Perspectives of Mathematics Education Group, June 1989.

Orton, A.   (1987)   *Learning Mathematics*, London: Cassell Education.

Quicke, J.   (1988)   *"The `New Right' and education"*,
British Journal of Educational Studies, 26, 1, pp.5-20.

Ruthven, K.   (1986)   "Differentiation in mathematics: a critique of
Mathematics Counts and Better Schools",
Cambridge Journal of Education, 16, 1, pp.41-45.

Scott-Hodgetts, R.   (1988)   "The National Curriculum: implications for the sociology of
mathematics classrooms", paper presented to Social
Perspectives of Mathematics Education Group,
December 1988.

Smithers, A. &
Hill, S.   (1989)   "Recruitment to physics and mathematics teaching: a
personality problem?",
Research Papers in Education, 4, 1, pp.3-21.

Straker, N.   (1987)   "Mathematics teacher shortages in secondary schools:
implications for mathematics departments",
Research Papers in Education, 2, 2, pp.126-152.

Tolkien, J.R.R.   (1937)   *The Hobbit*, London: George Allen & Unwin.

Whitty, G.   (1989)   "The New Right and the National Curriculum: state control or
market forces?",
Journal of Education Policy, 4, 4, pp.329-341.

Part 3

THEORETICAL FRAMEWORKS AND CURRENT ISSUES

INTRODUCTION

The four papers in this section examine issues in mathematics education from a philosophical perspective, by which is meant that they analyse some of the fundamental assumptions underlying these issues.

In her paper "Towards a Multi-Cultural Mathematics Curriculum", Marilyn Nickson examines the culture of the child from many angles, at the general level and at the level of the child, and points out the limitations of thinking just at any one level. She calls for a recognition of the social nature of mathematical knowledge, and an emphasis on the social aspects of the classroom, as compared to the individual psychological focus that has prevailed until now.

Paul Ernest's paper "The Nature of Mathematics: Towards a Social Constructivist Account" is a fore-runner of his recent book [Ernest 1990], and his eventual programme is a philosophy of mathematics education. In this paper, he sets up a social view of mathematical knowledge, and examines possible criticisms. Finally, he suggests consequences for mathematics education. He indicates the richness that comes from alternative perspectives of the nature of mathematics, for mathematics education, and highlights the importance of theory.

The theme of culture is picked up again by Leo Rogers in his "Mathematics Education and Social Epistemology: The Cultural History of Mathematics and the Dynamics of the Classroom". Roger's approach is to look at the social function of school mathematics, seen against the background of society's needs and expectations. He attempts to indicate some of the conflicting tensions and point to potential ways forward.

In the educational world in particular, we are aware of some of the overt and covert discouragements that affect the lives of girls. In her paper "Of course you *could* be an engineeer, dear, but wouldn't you *rather* be a nurse or a teacher or a secretary?", Zelda Isaacson takes this further and examines both the positive and negative inducements that affect girls. Mathematics is a crucial subject in the choices that students make, it being a necessary qualification for so many careers in which women are under-represented, as well as being seen as a measure of general intellectual ability, making Isaacson's analysis essential reading for mathematics educators in particular.

TOWARDS A MULTI-CULTURAL
MATHEMATICS CURRICULUM

Marilyn Nickson

THERE IS currently a degree of concern with the problems of how to meet the demands of educating children in a multi-ethnic society. At least one project is underway, for example, which is studying multi-ethnic secondary schools (Tomlinson 1986). However, in spite of such general concern, there is little evidence that the problem is receiving much attention from the standpoint of mathematical education in particular. Dawe (1983) has researched the effects of learning mathematics by children whose mother tongue is not English but other interest has tended to remain at somewhat general level.

In the course of working with a multi-ethnic group of middle-school pupils recently, the most striking feature to me has been the apparent sameness of the pupils rather than the differences. Coupled with this is the impact of their acceptance of this 'sameness', particularly where there views of classroom mathematics were concerned. Listening to the children's voices on tapes, one would be hard pressed to identify which of the children came from Afro-Caribbean, Asian or English background. The results might also challenge the preconceptions of many when identifying those who appear to be more able in thinking mathematically 'on their feet'. The impression gained is that the present tendency to cater for differences in pupils based on their ethnic group may be to approach the problem in unhelpfully restrictive terms and that we would do better to concentrate on the different cultural characteristics of pupils in the classrooms. Clearly, cultures are determined by ethnicity to a certain degree but then there are elements such as class, immediate physical environment (for example, living in a city or living in the country) and others which contribute to the complexity of what makes up a particular culture. D'Ambrosio's (1986) words help to clarify this when he writes:

> 'We have built a concept of society out of cultural attitudes and cultural diversity, that is, different groups of individuals behave in a similar way, because of their modes of thought, jargon, codes, interests, motivation, myths.' (p.5)

It will be argued here that to view the problem of developing mathematics curricula to cater for the needs of different ethnic groups may be to adopt a self-defeating approach at the outset because this is to see the problem from a perspective which is focussing too specifically on one aspect of culture. Rather, if we were to pay more heed to D'Ambrosio's (1986) notion of cultural grouping and, at the same time, to the social nature of mathematical knowledge, we would be better able to develop mathematics curricula which cater for cultural differences than is the case at the moment.

With this in mind, I would like to begin by briefly referring to the effects different views of mathematical knowledge have on the curriculum process and hence on classroom contexts. I shall then go on to consider other social aspects of mathematical knowledge and different cultures, and to relate these to the notion of a mathematical curriculum.

SOCIAL ASPECTS OF MATHEMATICS CURRICULA

A major problem in mathematics education is posed by the different perceptions of the subject that may be held by all of the participants in the process. There is a growing literature relating to the need for and identification of alternative theories relating to the nature of mathematical knowledge (for example, Confrey 1980, Lerman 1983, Nickson 1982, Pimm 1982, Wolfson 1981). Attention has been drawn to the potential effects that perceptions of the subject have on the social context of mathematics classrooms. (Nickson 1982). For example, a view of the subject from a formalist perspective described by Lakatos (1976) tends to produce a classroom situation of a somewhat sterile nature in which discussion, activity and investigation are the exception rather than the rule. Mathematics is very much received knowledge which is not to be challenged and generally, the pedagogy consists of chalk and talk followed by silent book work. This perception of the subject as infallible produces, in other words, a social context which emphasises the isolation of individual pupils, grappling silently with the mysteries of the subject. In short, it might best be described as 'anti-social'. Individual pupils are at the mercy of the subject which they may or may not succeed in understanding but if they do not, it is generally accepted that they cannot because of the abstruse, 'given' nature of the subject which only a few can aspire to understand. Lakatos (1976) gives us an example of an alternative approach within the mathematics classroom where a Platonic model is used, but more importantly, he points us away from the idea of the incorruptible formalist 'heaven' and suggests that mathematics need not be perceived in this way.

If we turn to other thinkers, we find that there exist epistemologies that provide the exciting possibility of mathematics being connected with, and indeed, founded in social activity (for example, Popper 1972, Kuhn 1970, Toulmin 1972). Clearly differences exist in the views of such philosophers but their essential importance when considered in connection with mathematics is that they encompass the notion of change and growth in knowledge within mathematics as within any other discipline. Mathematical knowledge no longer need be viewed as a 'special case' in this respect. Further, the growth and change within the subject has a social origin and takes place within a social context and not in an erudite vacuum.

If such a perspective of mathematics is adopted by educators, the implications for the classrooms in which it is taught are emphatic in their nature. Since the subject becomes more 'open', the criteria for the selection of content, the methodology and processes of evaluation in turn become more open. Pupils are encouraged to question, to theorise, actively to engage in testing theories and to explain how they view problems. The social context of the classroom becomes characterised by activity and discussion and inter-personal exchange between teacher and pupil, and pupil and pupil. The views of the individual pupil count and are shared and they are openly encouraged to express these views. This is not such a revolutionary approach to the subject for we read in 'Why, What and How? Some basic questions for mathematics teaching' published by the Mathematical Association (1976) the following aim for mathematics education: the stressing of 'mathematics as a social activity, in its conduct, its existence and its applications, with a concurrent emphasis on communication skills - verbal, graphical and written'. (p.2) Why is it so difficult to implement the kind of curriculum that would lead to the achievement of such an aim? It is arguable that when such an aim is pursued, the context of the mathematics classroom is better suited to cater for the different cultural aspects that individual pupils bring to the situation because of the give and take in the classroom context that arises. Perhaps somewhat paradoxically, by emphasising the social, the cultural differences of individuals are given space and allowed to enrich the mathematics curriculum. In order further to understand how this can come about, it is helpful to consider some of the social causes that can give rise to differences in mathematical thought.

SOCIAL CAUSES OF MATHEMATICAL THOUGHT

Bloor (1976) writes about what he calls 'an alternative mathematics' where he projects the notion of what mathematics might be like were we to develop mathematical knowledge unlike that which has been developed (pp.95-6). He suggests that 'A real alternative to our mathematics would have to lead us along paths where we were not spontaneously inclined to go.' (p.95) and compares this with the development of alternative moralities. He makes the point that 'the anthropologist will be prepared to talk of alternative moral systems provided only that they appear to be established and ingrained in the life of a culture' and acknowledges that this is the characteristic that must be identified in mathematics if 'alternative mathematics' were to make sense. What is of particular relevance to the present argument however, is his discussion of the effects of cultures upon the development of mathematical ideas. He states:

> 'The world does not, for the most part, consist of isolated cultures which develop autonomous moral and cognitive styles. There is cultural contact and diffusion. In as far as the world is socially blended then to that extent it will be cognitively and morally blended too. Again, mathematics like morality is designed to meet the requirements of men who hold a great deal in common in their physiology and their environment. So this too is a factor towards uniformity and a common backdrop of cognitive and moral style. Alterna-

tives in mathematics must be looked for within these
natural constraints.' (p.97)

This is helpful insofar as it recognises cultural difference but, at the same time, stresses the importance of commonalities. Bloor (1976) is concerned to find differences in mathematical thought which have arisen from social causes and he identifies five such causes:

(i) variation in the broad cognitive styles of mathematics;

(ii) variation in the framework of associations, relationships, uses, analogies, and the metaphysical implications attributed to mathematics;

(iii) variations in the meanings attached to computations and symbolic manipulations;

(iv) variation in rigour and the type of reasoning which is held to prove a conclusion;

(v) variation in the content and use of those basic operations of thought which are held to be self-evident logical truths.'

Bloor (1976) goes on to give historical examples of the first four of these and devotes and entire chapter to a discussion of the fifth. He shows how the development of aspects of mathematical thought have been linked through the centuries to social characteristics of prevailing cultures. I would like now briefly to explore how these social causes of mathematical thought may help in the notion of the development of a multicultural mathematics curriculum.

SOCIAL CAUSES OF MATHEMATICAL THOUGHT AND A MULTI-CULTURAL MATHEMATICS CURRICULUM

It was suggested earlier that by adopting an epistemological view of mathematical knowledge that stresses change, development, and its social foundations generally, and by consciously relating this to the curriculum process, the result would be to make the subject more open in its nature and more easily accessible. It was also suggested that this would lead to giving more attention to the thoughts and ideas of the individual pupil within the classroom. An examination of Bloor's (1976) social causes of mathematical thought identify particular factors that, if attended to within the mathematics curriculum, could act as a guidelines in catering for the cultural individuality of pupils. For example, different cognitive styles of pupils is a characteristic of which we would claim awareness as educators, and hopefully take into account in mathematical pedagogy. Those of us concerned with the history of mathematics would wish to enrich our teaching by reference to variation in the different symbolic systems used through the ages (and those in use in different cultures today) to emphasise the versatility that characterises mathematical thought. With respect to the notion of differences in rigour and what constitutes mathematical proof, Bloor (1976) refers to 'the varying standards of proof and logical discipline which have been felt appropriate at different times' (p.114) and gives examples of how in different places at differing periods of history, notions of rigour varied. In this present day and age, it may be that computers will influence our perceptions of rigour and proof in a way not previously met but certainly it is a tool to which pupils will turn to test ideas and explore their validity.

Clearly, it is the second of Bloor's social causes which could prove most fruitful in the context of our concern with and search for a multicultural mathematics curriculum. Probably more than any other factor, it is the framework of 'associations, relationships, uses, analogies, and the metaphysical implications attributed to mathematics' which would be most helpful in looking for ways to accommodate cultural diversity in the mathematics classroom and in particular ethnic diversity. Such a framework serves to draw our attention more closely to the individuality of the pupils and the different kinds of social interchange they experience. D'Ambrosio (1986) reflects this to some extent when he identifies two societal levels which pose concern for mathematics education. Firstly, there is the level of societal groups 'with clearly defined cultural roots, modes of production and property, class structure and conflicts, and senses of security and of individual rights'. (p.5) All children have this cultural aspect to their background, but they also have a societal group at the level of what D'Ambrosio (1986) refers to as 'children's societal arrangements' by which is meant the general social behaviour exhibited by children as a group. But D'Ambrosio (1986) goes on to point out that:

> 'Both have, as a result of the interaction of their individuals, developed practices, knowledge, and, in particular, jargons (the way they speak) and codes, which clearly encompass the way they mathematise, that is, the way they count, measure, relate and classify, and the way they infer.' (p.5)

This suggests that Bloor's (1976) social cause of mathematical thought relating to particular frameworks (referred to above) might best be viewed at both societal levels if we are to take into account cultural diversity amongst pupils in the classroom. There are many examples of variations in matters such as associations and uses of mathematics at the broader societal level, for example in the construction of boats by Amazon Indians to which D'Ambrosio (1986) refers, basket weaving in Mozambique (Gerdes 1985) and the use of computers in western society. Added to this, each society has its children's societal group within which, again, there exist a variety of frameworks. Within the latter, mathematics will once more have particular uses, relationships and so on. The most obvious examples would be those related to games that children play in which they mathematise at a primitive level, as in hopscotch where the shapes drawn on the ground must tesselate and are identified by numbers. Another example of a less spontaneous nature (insofar as it is contained within the mathematics curriculum) which comes to mind is the conceptualisation of mathematics in the context of LOGO where children experience and share mathematics in a way and at a level which has never happened before and from which we, as adults, are excluded (since this is an experience from which we did not benefit as children!)

CONCLUSION

Cultural diversity exists within both societal levels, at the general and at the level of the child. It is clear that a considerable tension must exist for pupils in mathematics classrooms between the two and a seesaw between which of the two predominates must continuously take place. As educators, we would all claim to know something about the cultural background at the level of the 'child' societal level, of those we teach, and how this aspect of their culture can affect the classroom mathematical learning situation. However, an awareness of the cultural background of each child at the broader societal level (and this is not solely a question of ethnic origin) is not a factor that we are inclined to attend to. If we were to take such considerations more into account and to couple them with an increased awareness of the ways in which mathematical knowledge can be socially determined, we might be better equipped to develop more effective mathematics curricula.

For details we have concentrated upon the psychology of individual differences and we have attempted to cater for these individual differences by attending to them in developing mathematics curricula. In order to develop a multicultural mathematics curriculum, however, a shift from our concern with differences of a psychological kind to those which are more social in their nature, and which will allow the cultural individuality of pupils to be accommodated, needs to be made. This paper has been concerned to present some 'first thoughts' about the possible direction such a shift could take. By attending more to the social origins of mathematical knowledge, exploring these 'social causes' and relating them to the cultural individuality of the children we teach, a multicultural mathematics curriculum could become more of a reality.

REFERENCES

Bloor D.   (1976)   *Knowledge and Social Imagery,*
                           London: Routledge & Kegan Paul Ltd

Confrey J.   (1980)   *Conceptual change analysis: Implications for mathematics and curriculum inquiry.* Paper presented to the AERA National Meeting,
                           Boston, Mass, February 1980

D'Ambrosio U.   (1986)   'Socio-Cultural Bases for Mathematics Education' in,
                           Proceedings of the Fifth International Congress on Mathematics Education,
                           Boston, Mass, Kirkhauser Ltd

Dawe L.   (1983)   'Bilingualism and Mathematical Reasoning in English as a Second Language', *Educational Studies in Mathematics 14, 325-353*

Gerdes P.   (1986)   'On culture: Mathematics and Curriculum Development in Mozambique', in Johnsen-Hoines M. and Mellin-Olsen S. (Eds) *Mathematics and Culture,* Bergen: Caspar Forlag, Bergen Laergerhogskole

Kuhn T.   (1970)   'Logic of Discovery or Psychology of Research' in Lakatos I. and Musgrave A. (Eds) *Criticism and the Growth of knowledge,* pp. 1-24 Cambridge: Cambridge University Press

Lakatos I.   (1976)   *Proofs and Refutations*
                           Cambridge: Cambridge University Press

Lerman S.   (1983)   'Problem-Solving or Knowledge-Centered: The Influence of Philosophy on Mathematics Teaching', *Inter-*

national Journal for Mathematical Education in Science and Technology, 13, 1 pp. 59-66

Nickson M.   (1981)   *Social Foundations of the Mathematics Curriculum: A Rationale for Change.* Unpublished Doctoral Dissertation: University of London, Institute of Education

Pimm D.   (1982)   'Why the History and Philosophy of Mathematics education Should Not be Rated X' in *For the Learning of Mathematics 3,1 12-25*

Wolfson P.   (1981)   'Philosophy Enters the Mathematics Classroom' in *For the Learning of Mathematics 2, 1, 22-26*

THE NATURE OF MATHEMATICS: TOWARDS A SOCIAL CONSTRUCTIVIST ACCOUNT

Paul Ernest

INTRODUCTION

I wish to apply two dichotomies to a discussion of the philosophy of mathematics: the prescriptive-descriptive distinction, and the process-product distinction. For much of this century, the philosophy of mathematics has focused on mathematical knowledge as a product, and has eschewed the process aspect of its epistemology. Knowledge as a product has been the focus of much of the philosophy of mathematics, not the process of coming to know. Secondly, the most visible accounts of the nature of mathematics and mathematical knowledge have been prescriptive - legislating how mathematics should be understood - rather than providing accurately descriptive accounts of the nature of mathematics. Although there are good historical reasons for these misconceptions, until they are overcome accounts of the nature of mathematics do violence to its reality. Once knowing, as well as knowledge is admitted as a legitimate concern of epistemology, it is possible to be more accurately descriptive. This leads to the acknowledgement of the human role in the creation of mathematics, that it is in fact a social construct. Once this view is adopted, it can be seen as relating to a number of other currents of contemporary thought. Such a case stands on its own merits, without a consideration of the educational implications. But once this case has been made, it can be seen to have immense educational implications. For a view of mathematics as a way of knowing and as a social construct can powerfully affect the aims, content, teaching approaches, implicit values, and assessment of the mathematics curriculum, and above all else, the beliefs and practices of the mathematics teacher.

THE PHILOSOPHY OF MATHEMATICS

The problem and task of the philosophy of mathematics is to account for the nature of mathematics. This is a special case of the task of epistemology, namely to account for human knowledge. The standard approach to this task is to assume:

Assumption 1 : There is a set of propositions that represents human knowledge in any field, and a set of procedures for verifying these propositions (or at least providing a warrant for their assertion).

Epistemology traditionally distinguishes between *a posteriori* and *a priori* knowledge, according to whether these procedures involve empirical verification or not (respectively). Mathematical knowledge is *a priori* knowledge because its propositions are established by means of logic, without recourse to empirical data.

Since the truths of mathematics are established irrespective of the facts of the world, no data can overturn them. Thus mathematical knowledge is the most certain of all forms of knowledge.

Traditionally the philosophy of mathematics has seen its task as providing a foundation for the certainty of mathematical knowledge. That is, providing a system into which mathematical knowledge can be cast, which provides a systematic way of establishing its truth.

Assumption 2: The role of the philosophy of mathematics is to provide a systematic foundation for mathematical knowledge, that is for mathematical truth.

This assumption can be illustrated with the well-known philosophical schools: logicism, formalism and intuitionism. Each of these schools utilises deductive logic as a means of warranting mathematical knowledge. Mathematical propositions which count as knowledge are either axioms, which are assumed or stand on their own merits, or the deductive consequences of axioms. Logicism rests its axioms on logic, intuitionism claims its axioms are self-evident, and formalism denies meaning or truth to its axioms at the object language level, but claims they are self-evident at the meta-language level. Each school justifies its rules of logic similarly. The schools are thus able to offer a programme for the warranting of mathematical knowledge, by re-casting it in their own way.

Ab initio the quest for certainty in mathematics is problematic, because of the precision that is needed in defining mathematical truth. We can distinguish between three truth-related concepts used in mathematics:

1. There is the truth of a mathematical statement relative to a background theory: the statement is satisfied by some interpretation or model of the theory.
2. There is the logical truth or validity of a mathematical statement relative to a background theory: the statement is satisfied by all interpretations or models of the theory.
3. There is the provability of a mathematical statement with the aid of assumptions drawn from a background theory: there is a finite logical proof of the statement from the axioms of the theory.

The second and third of these senses are what is usually meant by mathematical truth (the ambiguity matters little, since in most cases these two senses are demonstrably equivalent, by the appropriate completeness and soundness theorems). In a naïve sense truths are statements which accurately describe a state of affairs - a relationship - in some realm of discourse. In this view, the terms which express the truth name objects in the realm of discourse, and the statement as a whole describes a true state of affairs, the relationship that holds between the denotations of the terms. A mathematical truth in this sense is rare. Mathematical truths are made up of terms which usually do not name unique individuals, and the statement as a whole describes a structural relation which holds between whichever objects are named by its terms, in any appropriate interpretation (that is one satisfying the axioms of the background theory). This is uncontroversial. What it shows is that the concept of truth employed in mathematics no longer has the same meaning as either the everyday, naïve notion of truth, or its equivalent as was used in mathematics, in the past.

Whether we assume this new meaning of truth as it is employed in mathematics or not, it is still not possible to establish the certainty of mathematical knowledge. As Lakatos (1978) shows, despite all the foundational work and development of mathematical logic, the quest for certainty in mathematics leads inevitably to an infinite regress. Any mathematical system depends on a set of assumptions, and there is no way of escaping them. All we can do is to minimise them, to get a reduced set of axioms (and rules of proof). This reduced set can only be dispensed with by replacing it with assumptions of at least the same strength. Thus we cannot establish the certainty of mathematics without making assumptions, which therefore is not absolute certainty. Furthermore, if we want to establish that our mathematical systems are safe (i.e. consistent), for any but the simplest systems we are forced to expand the set of assumptions we make, further undermining the certainty of the foundations of mathematics.

What has been argued is that the traditional philosophies of mathematics which have assumed the task of trying to establish the certainty of mathematical knowledge have failed. However the point I wish to make is not only that such attempts are doomed, but that they are wrong-headed. Insofar as the philosophical schools of logicism, formalism and intuitionism offer an account of the nature of mathematics, it is prescriptive. That is, mathematics is a body of knowledge which should be seen as logic, intuitionistic mathematics, or consistent formal systems, according to which viewpoint is adopted. More generally, Assumption 2 is prescriptive, in that it requires us to focus on the systematic reconstruction of mathematical knowledge to warrant its assertion. It concerns how mathematics 'should be seen', to justify certain philosophical requirements.

The programmes and philosophies based on this assumption are akin to medieval chivalry or the Arthurian Legend. Namely, the pursuit of a lofty ideal - the Holy Grail - by an elite band with no concern for the mundanities of everyday life. Part of Cervantes' greatness is that his parody revealed this quest for what it was - a social construction of reality!

Should not an account of the nature of mathematics involve some consideration of how it is, as well as how it should be? A tenable philosophy of mathematics surely needs to be descriptive as well as prescriptive. However this question raises again the problem of what constitutes mathematics, and thus questions Assumption 1.

Is mathematics a body of knowledge expressible as a set of propositions, together with a set of logical procedures for their verification? It can be conceded that this is a part of mathematics. But evidently more can be considered. There are the processes of creating, transmitting and modifying such knowledge. Histories of mathematics document the creation and evolution of mathematical concepts and knowledge. Is this not also germane to the philosophy of mathematics?

There are a number of reasons for questioning Assumption 1, some general epistemological considerations, and some specific to mathematics.

138

First of all, since, as history illustrates, knowledge is perpetually in a state of change, epistemology must concern itself with the basis of knowing, as well as with the specific body of knowledge accepted at any one time. This is the view of pragmatists such as Dewey (1950), as well as modern philosophers of science such as Kuhn (1970) and Lakatos (1978).

Secondly, mathematical knowledge has an empirical basis, contrary to traditional views. This is the view put forward by an increasing number of philosophers, such as Lakatos (1976, 1978), Kitcher (1983) Tymoczko (1986). If this claim is accepted, then it is necessary to consider the process of coming to know in mathematics in considering the basis of mathematical knowledge.

Thirdly, if the task of the philosophy of mathematics is to account for the nature of mathematics, then other things being equal, a fuller descriptive account is to be preferred to a narrower account, provided it is coherent. For the parameters of the more extensive account allow it to better fulfil the task. Very powerful philosophical arguments would be needed to exclude whole areas of the domain in question from consideration at the outset of this inquiry, and I cannot find such arguments stated.

Given these three arguments, it seems appropriate to attempt to provide an account of mathematics that includes the process of coming to know as well as its product, that is knowledge. It is also appropriate to attempt to give a descriptive account, which locates mathematics where it is to be found in the world: in a social context. The next section offers a tentative social constructivist account of mathematics which attempts to overcome some of the difficulties described above.

MATHEMATICS AS A SOCIAL CONSTRUCTION

Philosophically, the social constructivist view of mathematics proposed here is both conventionalist and empiricist, in that human language, agreement and experience play a role in establishing its truths. Central to this view is the fact that over the course of time, mathematical knowledge changes, just as knowledge in the empirical sciences evolves. At any one given time, mathematics is an intersubjectively agreed, rather than an objective body of knowledge. It includes pragmatic rules governing procedures, as well as a body of propositions and methods.

It is easier to sketch such a view of mathematics in quasi-sociological terms. Thus a more suggestive, but tentative account along these lines is as follows. (For an account of social constructivism as a **philosophy** of mathematics see Ernest, 1990, 1991).

The starting point for any social constructivist account of mathematics is the assumption that the concepts, structures, methods, results and rules that make up mathematics are the inventions of humankind.

At any one time, the nature of mathematics is determined by three fuzzy sets, as well as the relationship between them. The three sets are: a set of information technology artefacts (books, papers, software, etc.); a set of persons (mathematicians); and more intangibly, a set of linguistically based rules adopted by, and activities carried out by, the mathematicians.

The set of mathematicians is partially ordered by the relations of power and status.

The informational content of the artefacts are the creations of the set of mathematicians, as they carry out the linguistically based activities constrained by the rules. This informational content becomes recognised as part of the concepts, methods and results that make up mathematics, when it is accepted by members of the set of mathematicians with power and status.

All the time these sets are changing, and thus mathematics is continuously evolving.

Versions of some of the concepts, methods and results that make up mathematics are represented in the minds of the mathematicians, and are symbolised in the information technology artefacts. Each of these representations includes a structure of relationships connecting the constituent objects and entities (the concepts). There is a powerful myth that these conceptual structures are all substructures of a single idealised structure of mathematics, or some past version of it. In fact, the various structures make up a family, with family resemblances between them (to use Wittgenstein's metaphor). In all likelihood, the intersection of all of the structures would be empty, even if we were to exclude instances of substructures that disagree with the idealised structure, and are considered incorrect. (However this operation of intersection is of course impossible, since the structures are intangible.)

Membership of the set of mathematicians results from the interaction of a person with other mathematicians, and with information artefacts. Over a period of time this results in a personal internalised mathematical structure and set of rules, derived through personal negotiation with other persons, until a 'fit' between structures is achieved (using the term in the sense of Glasersfeld, 1984). The set of mathematicians is a fuzzy set with different strengths of membership (which could be quantified from 0 to 1). This includes 'strong' members (active research mathematicians) and 'weak' members (teachers of mathematics). The 'weakest' members would simply be numerate citizens.

The set of rules and the idealised structure of mathematics are so significant that accounts of mathematics acceptable to many persons refer only to (some of) the rules and the structure.

The concepts of mathematics are derived by abstraction from direct experience of the physical world, from the generalisation and abstraction of previously constructed concepts, by negotiating meanings with others during discourse, or by some combination of these means. Many of the concepts generalised in mathematics come from the physical and other sciences (which likewise derive many of their concepts from mathematics).

Mathematics rests on spoken (and thought and read) natural language, and mathematical symbolism is a refinement and extension of written natural language. Mathematical concepts refine and abstract natural language concepts. Mathematical truths arise from the definitional truths of natural language, which is acquired by persons through social interaction. The truths of mathematics are defined by implicit social agreement on what constitute acceptable mathematical concepts, relationships between them, and methods of deriving new truths from old.

Mathematics is a branch of knowledge which is indissolubly connected with other knowledge, through the web of language. Language functions by facilitating the formulation of theories about social situations and physical reality. Dialogue with other persons and interactions with the physical world play a key role in refining these theories, which consequently are continually being revised to improve the 'fit'. As a part of the web of language, mathematics thus maintains contact with the theories describing social and physical reality. Of these, physical reality is the most obdurate, since there is an enduring physical world, even if our theories of it change. Through the physical sciences mathematics plays a key role in theories providing well-fitting descriptions of aspects of physical reality. Thus the 'fit' of mathematical structures in areas beyond mathematics is continuously being tested. Indeed, new mathematical structures invented to 'fit' other areas have been a source of renewal for mathematics throughout its history.

A key feature of mathematics is its perceived impersonality, its objectification. Thus the concepts, methods and other creations of mathematics are ruthlessly reformulated and altered by mathematicians, in contrast, say, to literary creations. Such changes are subject to very strict and general mathematical rules and values. This objectification of the rules in mathematics also has the result of offsetting some of the effect of sectional interests exercised by those with power in the community of mathematicians.

Mathematicians form a community with a mathematical culture, that is a more or less shared set of concepts and methods, a set of values and rules (which is often understood implicitly), within the contexts of social institutions and power relations.

This is a tentative and greatly simplified formulation of a social constructivist view of mathematics. It combines sociological, psychological, as well as philosophical perspectives. It also stands in need of justification. As an alternative, I anticipate some of the objections to this view and answer them.

OBJECTIONS TO THIS VIEW

1. If the most certain of all truths, namely mathematical truths, are but social constructs, then they are conceivably false. *A fortiori* there are no truths. Consequently the above view is not a true account, as there are none. Hence there is no need to accept it.

This is correct. The social constructivist view of mathematics is offered as a explanatory hypothesis, not as a truth. If made sufficiently precise, insofar as it is descriptive it could, in principle, be refuted. Failing that, its retention must depend on its explanatory utility.

2. The view presented above conflates philosophical, psychological and sociological explanations and concepts, and thus fails to offer a coherent account of mathematics from a philosophical perspective.

To insist that these fields must be kept separate is to prejudge the issue. For the above account makes the assumption that all human knowledge rests on human language, and hence is part of a connected web. There is a growing train of thought in philosophy which admits into epistemology considerations of human activity or pragmatics (Dewey, Wittgenstein, Polanyi, Toulmin), psychology (Piaget, Lorenz, Bruner, von Glasersfeld) or sociology (Kuhn, Lakatos, Feyerabend, Barnes, Bloor, Restivo). Thus a boundary-crossing approach has a good provenance.

Ultimately, the aim of this social constructivist account of mathematics, when fully developed, is to show that shared explanations can explain:

> the psychology of individuals learning mathematics;
> the historical development of mathematics;
> mathematics as a living social institution;
> the philosophy of mathematics.

This may be ambitious, but is not illegitimate. Currently modern physics is seeking to unify its various theories. The history of mathematics likewise provides plenty of examples of theoretical unification. This is also a worthwhile goal for the philosophy of mathematics.

These remarks notwithstanding, the account is not proposed as a purely philosophical account. For these, the reader is referred to Ernest (1990, 1991).

3. A social constructivist account of mathematics rests the 'certainty' of mathematical knowledge on shared language use, social agreement, and arbitrary definitions and conventions. It therefore cannot account for two central aspects of mathematics: the conviction held by many that it is objective and certain, and its "unreasonable effectiveness" (Wigner, 1960) in providing models of physical reality.

These are indeed two of the challenges facing this account. The certainty of mathematics is something that emerges gradually in learners of mathematics as they develop the mathematical part of language, and internalise the meanings of mathematical concepts and the relationships between them. Throughout this development there is interchange with others which leads to the agreement or 'fit' between individuals' constructions. (Although psychologists and mathematics education researchers well know that for many the 'fit' is poor.) Certainty only emerges as the end product of this constructive process. Thus the certainty of the equivalence between 'not-A or B' and 'A (materially) implies B' is unquestioned by most mathematicians and logicians. For many who have not been through the same learning experiences this will not be so. For they will not have internalised the refined mathematical meanings of the logical connectives, and their interrelationships. Thus the mathematicians' intuition of the certainty of this basic fact depends on their experience. Similarly '1+1=1' is only known to be false, with certainty, because of our learned definitions, rules and assumed interpretation. In fact it is true, with certainty, in Boolean algebra, when we routinely make different assumptions. The difference is that '1+1=2' is a truth of the mathematics embedded in natural language usage, whereas Boolean algebra is a more artificial creation (according to Russell, the first 'pure' mathematics).

It is the traditional philosophers of (pure) mathematics who have so much trouble with its "Unreasonable effectiveness". The social constructivist account by admitting the connection of knowledge, the experiential origins of mathematical concepts and structures, and the hypothetical nature of all knowledge finds more than coincidence in the applicability of mathematical structures to the world. However there is an important point here. What is proposed is not 'the social construction of reality', but the social construction of our knowledge of reality. And the process of this construction involves 'fitting' our conjectures as closely to reality as we can, with the possibility of falsification, as Popper (1959) suggests. Thus although our linguistic and scientific definitions and conventions are arbitrary, in that they have been freely created within the constraints of implicit or explicit rules, they have to withstand the gruelling tests of consistency not only within language, but also in 'fitting' of hypotheses to reality. (See Ernest, 1990, 1991, for a fuller development of these theses).

STRENGTHS OF THE SOCIAL CONSTRUCTIVIST ACCOUNT
The social constructivist account of mathematics is proposed because it appears fruitful in being able to provide satisfactory accounts of more aspects of mathematics than many other views. As is suggested above, it accounts for:

> mathematics descriptively instead of prescriptively,
> the growth and changing nature of mathematics,
> mathematical activity as well as mathematical knowledge,
> the links between the psychology, sociology and philosophy
> of mathematics, and
> the link between pure mathematics and the world.

In addition, the view can explain the apparent objective existence of mathematical objects, and the mechanism behind the creation and growth of mathematics.

4. How can the success of the platonist view, that mathematical objects and structures have an objective existence, be explained?

If the constructivist view that our knowledge of reality is a mental construction - albeit mediated by human interaction - is accepted, then not surprisingly, mentally constructed mathematical realities can be as potent as physical realities. Certainly other fictions have a powerful impact upon our lives. Consider only the concept of 'money'. We cannot deny the existence and power of this denotationless symbol. But it is clearly a social construct.

The uniformity of mathematical meanings amongst mathematicians, and a shared view of the structure of mathematics, result from an extended period of training in which students are indoctrinated with the 'standard' structure. This is achieved through common learning experiences and the use of key texts, such as Euclid, Van der Waerden, Bourbaki, Birkhoff and MacLane, Rudin, etc. (or their more recent equivalents). Most students fall away during this process. Those that remain have successfully negotiated and internalised a sub-structure that fits with a subset of the official one - at least in part. These mathematicians will have called up their mental mathematical schemas and used and reinforced them through use so much that they seem to exist objectively. Thus the objects of mathematics are mental constructions which have been given so much solidity, that they seem to have a life of their own. This accounts for why some of the greatest mathematicians are Platonists. They simply have made their mental worlds of mathematics more real than the rest of us by constantly revisiting and extending them.

5. What mechanisms account for the development of mathematical ideas?

Here Piaget's genetic epistemology has a great deal to offer. The mental constructions of individuals, through the processes of communication and negotiation, within the strict constraints imposed by the rules of mathematics (also mediated by the exercise of power by mathematicians) are added to the official conceptual structure of mathematics. Thus there is a social process by means of which the ideas of individuals spread out through enlarging groups of persons, and then may be taken up by 'official' mathematics. The actual genesis of mathematical ideas within individual minds involves vertical and horizontal processes (by analogy with inductive and deductive processes, respectively). The vertical processes involve generalisation, abstraction and reification, and are akin to concept formation. Typically, this process involves the transformation of properties, constructions, or collections of constructions into objects. Thus, for example, we can reconstruct the creation of the number concept, beginning with ordination. The ordinal number '5' is associated with the 5th member of a counting sequence, ranging over 5 objects. This becomes abstracted from the particular order of counting, and a generalisation '5', is applied as an adjective to the whole collection of 5 objects. The adjective '5' (applicable to a set), is reified into an object, '5', which is a noun, the name of a thing in itself. Later, the collection of such numbers is reified into the set 'number'. Psychologically, it is plausible that the objects of mathematics are 'reified constructions'. But this has also been offered as a philosophical account of the nature of mathematics by Machover (1983). Such a genetic account holds promise as a way of linking the psychology, history and philosophy of mathematics.

The horizontal process of object formation in mathematics is that described by Lakatos (1976), in his reconstruction of the evolution of the Euler formula and its justification. Namely, the reformulation of mathematical concepts or definitions to achieve consistency and coherence in their relationships within a broader context. This is essentially a process of elaboration and refinement, unlike the vertical process which lies behind 'objectification'.

This account, which suggests the mechanism which objectifies the concepts of mathematics, may also help to explain why mathematics as a whole is seen to be objectified. To project idealised structures into a platonic realm may be a natural application of this process of reification. It may be the case that mathematics would have been severely handicapped in its evolution without the myth of its objectivity, and the tendency to objectify it. For this myth permits a focus on the objects and processes of mathematics as shared things-in-themselves, subject only to the rules of the game.

The above sketch begins to show that a social account of mathematics is needed to do it justice, if the reconceptualization of the philosophy of mathematics offered above is accepted. Social constructivism is one of a number of possible accounts, and is further elaborated elsewhere (Ernest, 1991).

SOME IMPLICATIONS FOR EDUCATION

The present book is concerned with mathematics education, so it is appropriate to indicate, however briefly, some of the educational implications of the social constructivist account of mathematics.

One aspect of this view is that mathematics is seen as embedded in a cultural context. It leads to the conclusion that the view that mathematics somehow exists apart from everyday human affairs is a dangerous myth. It is dangerous, not only because it is philosophically unsound, but also it has damaging results in education. Thus, if mathematics is viewed as a body of infallible, objective knowledge, then mathematics bears no social responsibility. The underachievement of sectors of the population, such as women; the sense of cultural alienation from mathematics felt by many groups of students; the relationship of mathematics to human affairs such as the transmission of social and political values: its role in the distribution of wealth and power; the mathematical practices of the shops, streets, homes, and so on - all of this is irrelevant to mathematics.

On the other hand, if mathematics is viewed as a social construct, then the aims of teaching mathematics need to include the empowerment of learners to create their own mathematical knowledge; mathematics can be reshaped, at least in school, to give all groups more access to its concepts, and to the wealth and power its knowledge brings; the social contexts of the uses and practices of mathematics can no longer be legitimately pushed aside, the uses and implicit values of mathematics need to be squarely faced, and so on.

This second view of mathematics as a dynamically organised structure located in a social and cultural context, identifies it as a problem posing and solving activity. It is viewed as a process of inquiry and coming to know, a continually expanding field of human creation and invention, not a finished product. Such a dynamic problem solving view of mathematics embodied in the mathematics curriculum, and enacted by the teacher, has powerful classroom consequences. In terms of the aims of teaching mathematics the most radical of these consequences are to facilitate confident problem posing and solving; the active construction of understanding built on learners' own knowledge; and the exploration and autonomous pursuit of the learners' own interests.

If mathematics is understood to be a dynamic, living, cultural product, then this should also be reflected in the school curriculum. Thus mathematics needs to be studied in living contexts which are meaningful and relevant to the learners. Such contexts include the languages and cultures of the learners, their everyday lives, as well as their school-based experiences. If mathematics is to empower learners to become active and confident problem solvers, they need to experience a human mathematics which they can make their own. The social constructivist view places a great deal of emphasis on the social negotiation of meaning. Clearly this has very strong implications for discussion in the mathematics classroom.

The proposed view of mathematics legitimates ethnomathematics - the naïve, intuitive, pre-academic, culturally embedded conceptual structures and practices of mathematics. For these are seen to be identical to the sources of official mathematics. Scholars such as Bishop, D'Ambrosio and Wilder have been exploring the social origins of mathematics, on both the micro and macro scales, and these all acquire a new relevance for the classroom.

The social constructivist view also raises the importance of the study of the history of mathematics, not just as a token of the contribution of many cultures, but as a record of humankind's struggle - throughout time - to problematise situations and solve them mathematically - and to revise and improve previous solution attempts. By legitimating the social origins of mathematics, this view provides a rationale, as well as a foundation for a multicultural approach to mathematics.

CONCLUSION

This paper has attempted to sketch a social constructivist approach to the philosophy of mathematics. This is a first sketch, which is greatly simplified, and reference has been made to fuller accounts. Currently I am at work on a full scale treatment of Social Constructivism as a philosophy of mathematics (Ernest, forthcoming). The need for a social view of mathematics arises from the inadequacy of the traditional foundationist accounts. As I have indicated, those of us in education have an additional reason for wanting a more human account of the nature of mathematics. Anything else alienates and disempowers learners.

REFERENCES

Dewey, J.   (1950)   *Reconstruction in Philosophy,*
                               New York: Mentor Books.

Ernest, P.   (19901)   Social Constructivism as a Philosophy of Mathemat-
                               ics: Radical Constructivism Rehabilitated? Presented
                               at *PME-14,* Mexico, July 15-20, 1990.

Ernest, P.   (1991)   *The Philosophy of Mathematics Education,*
                               London: Falmer Press.

Ernest, P.(forthcoming)   *Social Constructivism as a Philosophy of*
                               Mathematics,
                               Albany, NY.: SUNY Press.

Glasersfeld,
E. von   (1984)   'An Introduction to Radical Constructivism', in P.
                               Watzlawick, Ed., *The Invented Reality,*
                               New York: Norton, 1984, pp. 17-40.

Lakatos, I.   (1976)   *Proofs and Refutations,*
                               Cambridge: Cambridge University Press.

Lakatos, I.   (1978)   *Philosophical Papers* (Volume 2),
                               Cambridge: Cambridge University Press.

Kitcher, P.   (1983)   *The Nature of Mathematical Knowledge,*
                               Oxford: Oxford University Press.

Kuhn, T.S.   (1970)   *The Structure of Scientific Revolutions,*
                               Chicago: Chicago University Press.

Machover, M.   (1983)   'Towards a New Philosophy of Mathematics',
                           British Journal for the Philosophy of Science,
                           Volume 34, pp. 1-11.

Popper, K.   (1959)   *The Logic of Scientific Discovery,*
                           London: Hutchinson.

Tymoczko, T.
(Ed.)   (1986)   *New Directions in the Philosophy of Mathematics,* Boston:
                           Birkhauser.

Wigner, E.P.   (1960)   'The unreasonable effectiveness of mathematics in the
                           physical sciences', reprinted in T. L. Saaty and F.J.
                           Weyl Eds. *The Spirit and Uses of the Mathematical
                           Sciences,* New York: McGraw-Hill, 1969, pp. 123-140.

EDUCATION AND SOCIAL EPISTEMOLOGY:
THE CULTURAL ORIGINS OF MATHEMATICS AND THE DYNAMICS OF THE CLASSROOM.

Leo Rogers

Revised September 1990

The following paper is an attempt to consider the mathematics classroom from the point of view of social epistemology, where the context of mathematics learning and teaching is influenced by the individual backgrounds of pupils and teachers, and by the external pressures and assumptions of institutions and organisations.

It is one thing to identify school mathematics, but quite another to specify the role it plays in the general school curriculum, and in the wider social frameworks in existence today. (1)

SOCIAL EPISTEMOLOGY
If we take a maxim from anthropology, that the current use of a tool does not necessarily indicate its origin, and consider the following transposition; that the current form of knowledge does not necessarily indicate its intended use, we may be led to consider the following questions:

What is the intended present use of school mathematics?

and

What are the exact social functions of school mathematics?

We may think we gain the original intended use from "tradition", and from declared intentions in the prefaces of books, publishers' advertising, various official and unofficial policy statements and so on. However, we should be wary of such assumptions.

It is generally taken for granted that the current content of school mathematics is "useful", and that this usefulness justifies its existence and continued transmission to generations of pupils. The social pressures are so great and the assumptions so pervasive that few are able to question this, and circumstances rarely arise so that the assumptions may be seriously challenged.

In the last thirty years or so, mathematics education has begun to identify itself as a distinct area of study and consider the following broad areas as an interrelated complex:

a)   New ideas about the nature of learning implying that the ability or otherwise of pupils to learn is seen as a function of the **methodology of the teacher** rather than a failing in the mental capacity of the pupil. Also, frequently neither the appropriateness of the mathematical content nor the ideology supporting it are questioned.

b)   The appropriateness of the mathematical content is being questioned in relation to the varied ideological standpoints of a variety of authorities, both local and national.

c)   New technology is introduced into the classroom, which may force non-traditional situations upon the established methodology. (For example the introduction of calculators allows pupils to discover decimals or negative numbers "too early"). (2)

d)   New economic needs arise which motivate the introduction of new content, forcing out the old or leading to syllabus packing.

e)   Competing methodological paradigms arise, thus enabling content to be seen from a new perspective.

f)   Fundamental changes in philosophical attitudes towards the subject matter of mathematics itself imply radical changes in methodology and in content. The shift of focus here has been away from the technicalities of so-called foundational issues and towards the social and epistemological contexts of the creation of mathematics in the individual and the group.

These aspects have reacted upon each other, each being championed by some individual or group, each being used for different public and private purposes. For broad social aims to be achieved, a commonality of purpose is necessary, but within the institutionalised setup, social mechanisms for distributing knowledge do not themselves function in a sufficiently invariant manner to ensure that truths will be transmitted intact. (3) Some of the worst examples are the "cascade method" for in-service training, where advisers shut in an upper room with an appointed prophet are sent forth to convert teachers to the new methodology. Even the old established method of teacher training itself works on the belief that knowledge can be packaged and passed on to pupils. These methods are doomed to failure, or at least only partial success, because the transmission of knowledge is **not systematically regulated.** While there may be a hierarchy of **function** there is no hierarchy of **control.**

This suggests that there is a collection of incommensurable ideologies of mathematics and its teaching, all roughly purporting to be about the same thing, but each one showing significant differences from the others.

REGULATING ACTIVITIES

If we hope that some common approach is possible, we may be led to ask what social mechanisms provide a regulating effect on this complex. Public examinations are intended principally to regulate content, and more recently to some degree, methodology. Examination technique implies choices of content and learning styles. Recent innovations in assessment imply that methodology can also be regulated. (4) Published materials again attempt to regulate, and for the majority of teachers examinations and published course materials represent the two major influences on their attitudes to teaching mathematics. Local meetings of teachers for various purposes may be intended to regulate, but can also produce a severely disruptive effect on the system by emphasising anomalies and by forcing changes. In this way we demonstrate that different social groups have incommensurable aims for mathematics in school, due to their radically different ways of structuring experience.

The world-view of the current political power group is based on a particular ideology of competitive capitalist economics. Aspects of this ideology require the vocational training of the potential work force in order to provide skilled manpower to preserve and maintain the economic stability of the nation. While this general aim may be worthy, the means are in dispute. Particular comparisons are drawn between our nation and its "competitiors" in order to establish a need for the reorganisation of the means of training the majority of pupils; the state school system.

"We must raise standards consistently, and at least as quickly as they are rising in competitor countries. The government now wishes to move ahead at a faster pace to ensure that this happens and to secure for all pupils in the maintained schools a curriculum which equips them with the knowledge, skills and understanding that they need for adult life and employment". (5)

The Cockcroft report insists that the study of mathematics by children at school is essential, but why? Mathematics is important for normal life (whatever that may be), and there is a concern that children should do well. Mathematics is useful at all sorts of levels, but the impact of technology promotes the view that mathematics is the magic posessed by a few individuals. Mathematics is something to enjoy, but the quality of its intrinsic delight evades most pupils. Mathematics is part of our cultural heritage, but rarely do we find any real attention paid to this aspect. There is no general consensus from society on what it wants from mathematics education for the average pupil, and this public vacuum has been filled by political initiative. (6)

Aims for mathematics education currently move away from the hitherto restricted list into wider areas of "higher level" cognitive skills, analysis, synthesis, evaluation, and the affective areas of receiving, responding, valuing and organising. (7) Communication is a vital aspect. The method of proofs and refutations becomes the paradigm cited to show that mathematics grows by social interaction. Choices are made, concepts are refined, procedures are agreed by criteria which ultimately depend upon **aesthetics** (the elegance of a proof, or the coherence of a perceived pattern), **economics** (what is required, or what is physically possible), **psychology** (what the participants of the discussion **perceive** to be necessary or pleasing), and fashion (what are the currently accepted themes, proofs and procedures).

We may cite some examples of institutional reforms in order to examine their impact on the way in which mathematics is viewed by different social groups. Napoleon's reform of the Ecole Polytechnique was a disaster in terms of the original intentions of the founders, and had some unforeseeable effects. (8) Impressed by the savants conscripted for the Egyptian expedition, he instituted reforms which required army officers, particularly the engineers and the gunners, to be trained in mathematics. Civil servants responsible for the development of roads and canals were also included. Defeat of the Prussians by Napoleon motivated great changes in the German school curriculum. Their pride hurt by superior French military skills and organisation, a considerable amount of extra mathematics both in content and time, was introduced in the belief that this would provide the necessary background for military and economic success. (9) It is highly unlikely that the mathematical training of Prussian youth contributed to Napoleon's defeat at Waterloo, but it is considered that the significant number of eminent German mathematicians of the latter nineteenth and early twentieth century is a direct consequence of the reforms. Other European countries followed suit and central-ised their education system in the belief that similar success would follow.

In England a differen pattern emerges. The industrial revolution showed workers that it was important to be educated. (10) To read, to write, to calculate and to learn to apply knowledge became vitally important to secular society and the workers' movements. Academics, colleges and other instititions were founded, where general education could be obtained. The demand for education was met by many enterprising individuals and private institutions. Gradually, during the nineteenth century, these became organised on first a local, and then on a national basis. (11)

The content of the mathematics curriculum, in contrast to the central direction of some continental countries, became a collection of "useful knowledge", propa-gated through minute books, handbooks, broadsheets, serials and various text-books. (12) Individual schools and colleges had considerable independence to decide what path to take, what curricula to establish, and to choose what mathemat-ics appeared to be useful for them at the time. The subsequent amalgamation of institutions and their syllabuses produced a collection of topics of "interpreted" mathematics which became the basis of the mathematical education provided in the technical colleges towards the end of the century.

A similar conglomeration of useful topics for the elementary schools was provided by the local school boards and the education societies with their hand-books and arithmetic texts, and much later by the subject associations. By the early twentieth century the Mathematical Association had begun to produce a definitive syllabus for the secondary (grammar) schools, and content and methodology had undergone some critical appraisal at the British Association by "outsiders" like Perry and Heaviside, speaking from the point of view of consumers of the "interpreted" mathematics. (13)

If we regard the textbooks, syllabi, curricula and examination papers of contemporary school mathematics, a naïve view might lead us to believe that this is sufficient to identify a "standardised" mathematics, general enough to be internationally recognisable, but subject to local variations dependent on national education systems and local demands, traditions and influences. (14) In view of the statement of the basic problems of epistemology, we cannot assume even this, for there is **no standardisation of interpretation of texts or directives.** Even given the attempted standardisation of the interpretation of content by examination requirements, we cannot guarantee that pupils even approach the same problems in the same sort of way. The fact that there appears to be a consensus of opinion borne out by the fact that pupils pass examinations does not really say much about the concepts they possess not the methods they employ.

Experience of any of us as examiners shows that even (to our mind) our easiest questions get a very varied response.

The examination attempts to test the student's knowledge (facts, skills, processes, etc.), and the examiners have particular and idiosyncratic views of what they want to test, how to do it, and what is acceptable as an answer. (The current disparity of views among the examining boards over GCSE coursework is a case in point).

The acceptable answer and the manner of its getting reflects the world-view of the examiner; in particular, the examiner attempts to extract from the candidate what aspects of their own world view are possessed by the candidate. Clearly, there is no **guarantee** of commonality of purpose or result. Formally, even the technical result of getting the answer does not say anything about **how** it was obtained, nor what view of mathematics the student holds.

With the broadening of views about what may be assessed, and how this might be done, we are attempting to provide a wider range of acceptable responses, and thereby complicating the issues even further.

COLLOQUIAL MATHEMATICS

The mathematics which naturally lies within human culture has its rooots in the diverse human activities carried out by all people; classification, appreciation of relationships, measuring, weighing, estimating, recognition of spatial forms, etc., and the implicit understanding of concepts and methods are what we seem to need for "everyday life". (15), (16) This is integrated with our language and other forms of symbolism into our total experience. We know children in a natural and unconscious way have already acquired a fair amount of informal mathematics before they come to school. (17) Mathematical ideas are often undifferentiated from the human activity like playing snooker, canoe-building or shopping, and appear in these to a greater or lesser degree. There is a clear distinction between the mathematical description or model of an activity, and the "natural" or informal mathematics required to carry it out. One of the severe difficulties with the pure utilitarian approach is the attempt to make explicit precisely what mathematics lies in these human activities. This results in the classical dissection of the activity into so many separate facets, and a resultant trivialisation of the mathematics and even the activity itself. (18) Take, for example, a game of netball. Where is the mathematics here? It would be very hard to argue that it actually exists in the game or the awareness of the players at all, while the game is being played, and it is difficult to see that there is any way in which we can realistically identify any formal mathematics, even at an elementary level in this situation.

However, some analysis of the game gives us information at a pre-mathematical level; the estimation of heights and distances for shooting and passing, which is part of the personal body experience of each individual and the essential skills of the game; and a more formal level which is to do with the tactics and strategy of the conduct of the game itself (which here includes the dimensions of the pitch, the height of the net, and the rules of play). While these aspects are functions of the instantaneous decision-making of the players themselves, they can be extracted from the game, and examined in terms of the pure mathematics involved. But then the mathematics detached from the real experience becomes an abstract intellectual activity. So many of the approaches to school mathematics miss the essential point that too early formalism actually takes away the essential contexts and reduces the experience to "trivial pursuits". The basic tenet that "Colloquial mathematics can and should serve as the starting point of mathematical education at school" (19) is sadly neglected.

ACADEMIC MATHEMATICS

If Colloquial mathematics is a term used to describe a considerable amount of mathematical activity that goes on in school, then we might claim that Academic mathematics is the intended outcome of the formalisation of the former. As far as University mathematics is concerned, the popular image is of institutionalised Bourbakist mathematics where the main emphasis is on research, and the criteria for judging "good" or "bad" mathematics are internal. Structures are built on interrelationships of mathematical concepts, their logical relations, and the establishment of theorems by proofs. However, we know that the **creation** of such mathematics is an **heuristic process,** and the **justification** is a **social** one. (20) In this case, the textbook serves as an introduction to the technical processes, but this may very well cloud the issues and mislead the student unless **interpreted** by a teacher.

In many cases, particularly at research level, we are still "learning at the feet of the master" in the same way that the Bernoullis went to Leibniz, or the American graduates went to Hilbert. The master demonstrates to the pupil certain rules of the art which may be useful, but only serve as a guide and need to integrated into the practice of the art. "To learn by example is to submit to authority. You follow your master because you trust his manner of doing things even when you cannot analyse and account in detail for its effectiveness. By watching the master and emulating his efforts in the presence of his example, the apprentice unconsciously picks up the rules of the art **including those which are not explicitly known by the master himself"**. (21) Thus, the **meanings** of Academic mathematics are transmitted largely by social activity, and are **not context-free** as popularly supposed. The aim of much school mathematics is to imitate this Academic mathematics. For example, Dienes' "six stages", (22) a model popular with teacher trainers, sets up a scheme for the final achievement of axioms, proofs and theorems. Successful achievement of this aim results in sterilised mathematics with the dialectic removed.

INTERPRETED MATHEMATICS

Interpreted mathematics is the context-bound use of mathematics as a tool, a means to an end, to solve problems in the "real" world. Interpreted mathematics is often very close to colloquial mathematics in the sense that within a subject area, biology or engineering, say, a kind of folk-lore about what are useful or appropriate methods is set up. The art of application of these methods is taught, as before, by a "master".

Ther is a standard level of elementary mathematical techniques which infiltrate the general training, with another level of newly-tried mathematical ideas which come from contacts with other sciences or academic mathematics. The adoption and application of new ideas and techniques is only possible by the kind of social intercourse outline above. Recent examples of this infiltration are the applications that have found their way into geography at school level, (23) and theoretical biology (24). What is actually happening here is a move towards what Agassi describes as learning by agenda, where "learning by agenda is ever so much more powerful than learning by textbook .... the increased efficiency is twofold. First, specificity. The mathematics required by (1) the amateur, (2) the applied mathematician, (3) the mathematics teacher, and (4) the research mathematician, are so very different that each needs a different agenda. Even when all four want to know what an axiom is, each of them will doubtless approach matters differently. And agenda-making is active student participation, and educational psychology is unequivocal about certain matters: there is no better training than by active participation ... writers in different fields collude: Weiner in cybernetics, Piaget in developmental psychology, Chomsky in psycholinguistics, Popper in scientific method; they all favour active participation". (25)

THE DYNAMICS OF THE CLASSROOM

We would not now question the significant and subtle role of language in the mathematics classroom; however, language through its spoken and printed forms is not the only vehicle for communication, for the whole range of actions, impressions, expressions and visual display is brought into use in any communicative act. Attempts to delete a large part of this wide spectrum of communication in the transmission of "precise" mathematics lead to sterility, misunderstanding and total lack of comprehension. Mathematics is transmitted, but never discussed. We might suggest that many teachers hold an ideology of mathematics and its teaching which is a *priori* with regard to **facts**, **utilitarian** with regard to **processes**, and **infallible** with regard to **truth.** The Lakatosian model does more than insist that the social dimension is vital to creative mathematics learning, it produces an awareness of deeper issues, principally the fact that the nature of knowledge is contingent upon human communication. I have argued elsewhere that it is only possible to arrive at what is here called academic mathematics by a necessary progression through other aspects of mathematical activity. (26) These aspects are exemplified in Popper's Three Worlds where, in learning and teaching mathematics we are concerned with:

a)   Physical objects, physical and mental states and potential proble.n-situations recognised and acted upon by individuals.

b)   Creative processes, belief-systems, practitioners' maxims, and the transformations upon these by way of heuristic.

c)   Mathematical structures, and the formal processes for the production and practice of mathematics. (27)

While the motivating situations lie in the first world, the bulk of colloquial and interpreted mathematics refered to above lie in the second world. Thus children faced with, say, Cuisenaire rods may apprehend them in the first world sense, but until motivated to act, do not develop their own knowledge in a subjective manner by the use of their creative powers. A long time is spent in this phase of contemplation before it is even possible to begin to objectivise facts, theories or arguments and separate them from the necessary aspects of individual personal knowing. Recognition of this social construction where the "objects" are mental states is clear where Bishop "views mathematics classroom teaching as **controlling the organisation and dynamics of the classroom for the purposes of sharing and developing mathematical meaning".** (28)

So far, the attempts at regulating the activity of the classroom in any positive manner, good or bad, by external means seem doomed to failure. The principal reason for this is that the regulators do not perceive the true nature of the experience nor the status of the knowledge generated. Ironically this may have saved us from some disasters, but the experience we have leads us to believe that the current situation is largely the result of survival tactics on the part of the teachers.

The master instructs the students in the art of mathematics. The particular words, expressions and nuances form the rhetoric which carry the objects of our experience. The heuristic method involves the individuals concerned in the dialectical construction of personal knowledge. These aspects are integrated into one dynamical system where students and teacher together become aware of the power of their own mathematical abilities and creations. Set against this, we have the principal role of the school to maintain the social system and preserve the status quo with regard to the nature of mathematical knowledge. How can we find ways of changing the situation so that the reality may be perceived?

REFERENCES AND NOTES

(1)   For a statement of purpose of Social Epistemology see:
              Fuller, S *Social Epistemology* 1 (1) 1987 (1-4)

(2)   See CAN Continuation Project, Homerton College,
              Cambridge CB2 2PH

(3)   See Fuller 1987 p2

(4)   Various taxonomies of educational objectives exist,
              from e.g. Bloom and Avital and Shettleworth through
              to the current APU scheme

(5)   DES *National Curriculum 5-16 July 1987,* pages 2-3

(6)   See Howson & Wilson, *School Mathematics in the 1990s*
              C.U.P. 1986, in particular, chapters 2,3, and 4

(7)   See Dorfler & McLone, *Mathematics as a School Subject*
              in Christiansen, Howson and Otte, *Perspectives on*
              Mathematics Education Reidel 1986 (55-58)

(8)   Bradley, Margaret. *Scientific Education Versus Military*
              Training: The Influence of Napoleon Bonaparte on the
              Ecole Polytechnique *Annals of Science* 32 (1975) 415-449

(9)   Gerstell, M Prussian Education and Mathematics.
              Amer. Math. Monthly 82 (3) 197 (240-245)

(10)   Simon, B. *The Two Nations and the Educational Structure*
              1780-1870 London 1974

(11)   Inkester, I. *Science and the Mechanics Institutes*, 1820-
         1850: the Case of Sheffield. *Annals of Science* 32 (1975),
         451-474

(12)   Simon (1974)

(13)   Perry, J. *The Teaching of Mathematics* British Associa-
         tion meeting, Glasgow, 1901. in Bidwell and Clason
         Readings in the History of Mathematics Education NCTM
         Washington 1970 (220-245)

(14)   See Dorfler & McLone 1986

(15)   See Dorfler & McLone 1986

(16)   Wilder, R.L. *Mathematics as a Cultural System* Pergamon
         1981 chapters 1 and 2

(17)   Many examples can be found in Ginsburg, H. *Chil-
         dren's Arithmetic* Van Nostrand, 1977, and others

(18)   The Classical and Romantic analyses of experience are
         described in Prisig, R.M. *Zen and the Art of Motorcycle
         Maintenance* Bodley Head 1974, in particular, chapters
         8-11

(19)   Dorfler & McLone 1986 p59

(20)   See Lakatos, I. *Proofs and Refutations* C.U.P. 1976

(21)   Polyani, M. *Personal Knowledge* Harper and Row 1964
         p53

(22)   Dienes, Z.P. *The Six Stages of Learning Mathematics*
         NFER 1972

(23)   See, for example: Cole and Beynon, *New Ways in
         Geography* Blackwell, 1968

(24)   Thom, R. *Structural Stability and Morphogenesis*
         Benjamin, 1976

(25)   Agassi, J. Mathematics Education: The Lakatosian
         Revolution in *For The Learning of Mathematics* 1 (1) 1980
         (27-319)

(26)   Rogers, L. The Philosophy of Mathematics and the Methodology of Mathematics Teaching: Zentralblatt fur Didaktik der Mathematik, *Analysen*, 2, 1978 (63-67)

(27)   Popper, K. *Objective Knowledge, An Evolutionary Approach* O.U.P. 1972

(28)   Bishop, A. The Social Construction of Meaning - a Significant Development for Mathematics Education? in *For the Learning of Mathematics* 5 (1) February 1985 (24-28)

OF COURSE YOU COULD BE AN ENGINEER, DEAR, BUT WOULDN'T YOU RATHER BE A NURSE OR TEACHER OR SECRETARY?

Zelda Isaacson

That fewer girls than boys engage in mathematics at school and beyond, and that girls on average perform less well than boys, especially at the higher levels of achievement, is well documented (e.g. Burton (1986), Isaacson (1982), The Royal Society (1986)). These facts are, increasingly, a cause of concern in the U.K. and indeed internationally (Note 1). In recent years the flow of literature on the 'gender and mathematics' issue has substantially increased, and much of this writing has been concerned with possible explanations of the phenomenon of female underachievement.

In this paper I wish to put before readers two theoretical constructs, namely, **coercive inducements** and **double conformity**, which individually, and even more together, offer powerful explanations and far-reaching insights and are therefore potentially of immense value to researchers in this field.

The notion of a coercive inducement is one which Helen Freeman and I developed some years ago. It seemed to us that to suggest that girls were *prevented* from taking up non-traditional roles in society - or 'not allowed' to engage in "boys'" subjects at school and higher education levels - was far from the felt experiences of most girls and women. Experientially, most girls do not see themselves as *forced* into submissive roles, or *coerced* into taking up low status and often subservient jobs - or into studying Child Development rather than Physics at age 14+. Rather, these female or feminine roles, jobs, school subjects, etc. are *chosen* by them, but they are 'chosen' because of a system of rewards and approvals which act as inducements and which are so powerful that they come to be a kind of coercion.

Observing young girls at play and at work in infant school classrooms, (Note 2) these patterns are clear. Girls regularly choose to play in the Wendy House (it may officially be called the 'domestic play area', but children and teachers alike slip into the familiar name!), they do not often choose to play with Lego or other constructional toys; they choose pretty clothes, Care Bears or My Little Pony for their birthday presents; six and seven year old girls often see their adult lives in terms of getting married and having children. If asked whether they will have a job when they grow up, even academically able girls name such things as, of course, nursing or teaching - or being a secretary or working in a sweet shop. It is crucially important, however, to acknowledge that the choices girls make day by day and hour by hour are, on the face of it, very attractive ones, and that the intrinsic and extrinsic rewards (e.g. personal satisfaction, the approval of significant others) accruing to girls for appropriately feminine behaviour (caring for others, helping others, building up relationships with others) are a further reinforcement of these patterns.

As Helen Freeman and I wrote in a draft paper when we were developing the notion of a coercive inducement a few years ago:

> *It has been suggested that a submissive role is forced upon girls through punishment of non-conformist behaviour. It seems to us, however, that it would be closer to the truth to suggest that, rather than being coerced into 'feminine' behaviour, girls are induced by a system of rewards and approval to accept a more passive role.*

And in a recent paper (Isaacson, 1986) I expressed the idea in this way:

> *There is a sense, I believe, in which many girls are persuaded to adopt typically female modes of behaviour and to choose stereotypically feminine occupations and life styles because the rewards for 'feminine' behaviour are too great to be refused, rather than because they are prevented from choosing others...*

Earlier this year (1987), in a *That's Life* programme on television, viewers were shown a small girl taking part, with an adult male partner, in a ballroom dancing display. The rewards she obtained for this were enormous - applause, lots of attention, being treated as 'grown-up' and being dressed up in a miniature version of the bespangled costumes of the adult women dancers. Very few children would be able to resist such delights. Certainly, this little girl revelled in her role. The inducement to play the feminine role was so great that it, in a very real sense, could not be refused by that child and so was a form of coercion - an offer which could not be refused. All the little girls who watched that programme, too, were being exposed to this coercion. The message came through loud and clear 'be a real woman, wear pretty, spangled clothes and dance on a strong man's arm, and you too will be applauded and fêted'.

What has any of this to do with mathematics learning? Very simply, it is that the learning of mathematics cannot be divorced from the social context in which that learning takes place. Mathematics (and science and technology) carry strong 'male' images, partly because they are seen as 'hard' - not necessarily intellectually difficult, but hard as opposed to soft (or feminine, yielding etc.). The male image of these subjects has a long history, and is further reinforced both by the fact that they have traditionally been dominated by men, and that their public concerns are typically masculine concerns, such as warfare, machinery and work aimed at subduing or controlling nature. (See, for example, Easlea 1981, 1983.) Girls and women often come to believe, therefore, that these 'masculine' subjects, and the jobs they lead to, are not for them. Worse, they often believe (at a subliminal level even if this is not made explicit) that if they engage in these activities they will put their valuable femininity (valuable because of the rewards it brings) at risk.

The effect on mathematics learning is cumulative. At option choice time, girls are often reluctant to continue with study of the physical sciences and technical subjects: because of the male image they carry; because of a relative lack of experience in these subjects which goes back to infant school days; and because they are making positive choices for 'feminine' subjects (human biology, needlecraft, art). Mathematics, then, becomes increasingly unpopular amongst adolescent girls, not only because in itself it carries an offputting male image (reinforced by boys in mixed sex classrooms who claim the subject as their own) but also because currently one of the most compelling reasons for pupils to learn mathematics - i.e. that it will be useful for other subjects studied and for future careers - is also absent for the girl who has already opted out of science and/or technology and has decided that that sort of job is not for her. There is research which suggests that pupils who study physics or technical subjects alongside O level mathematics, do better at mathematics than those who do not (Sharma & Meighan, 1980). So, although mathematics is a compulsory subject for the vast majority of pupils to age 16 in the U.K., girls often disengage themselves from the activities of the mathematics classroom. They are generally not permitted to give up mathematics altogether, but they can and do drop down and achieve far less than might have been predicted for them at an earlier stage of their schooling.

I grant that the system of rewards and approval for feminine behaviour which I said act as 'coercive inducements' would not normally be called coercion. The coerced person is usually understood to be doing what they are unwilling to do, because of fear of unpleasant consequences ('your money or your life'). The coerced person is thus normally understood to be unfree, whereas acting in response to an inducement is usually regarded as acting freely. I wish to argue, however, that this juxtaposition of the two concepts offers a way into understanding the mechanisms of female underachievement. Girls and women who 'choose' the path of conventional femininity are in one sense acting freely - they could have chosen otherwise - but in another sense are unfree. When the rewards for being a 'proper woman' are huge, while by not conforming one risks the loss of these rewards, then one's freedom is no more real than the freedom of a person living below the poverty line to take an expensive holiday abroad.

The notion of a coercive inducement on its own, I suggest, goes a long way to explaining why females are underrepresented in 'male' subjects and occupations. However, when combined with the second construct under consideration in this paper, that is, **double conformity**, its explanatory force is greatly increased.

I am indebted to Sara Delamont for the latter idea. In an essay entitled 'The Contradictions in Ladies' Education' (1978) she claimed that:

> The central theme which can be traced through the estab-
> lishment of education for middle and upper class girls and
> women from the 1840's to the present day is double
> conformity. This double conformity - a double bind or catch
> 22 - concerns strict adherence on the part of both educators

> *and educated to two sets of rigid standards: those of ladylike*
> *behaviour at all times and those of the dominant male*
> *cultural and educational system.* (p. 140)

Double conformity expresses the dilemma of any person who is in a situation where they have to conform, at the same time, to two sets of standards or expectations, where these two sets are mutually inconsistent. This was the case for the pioneers of women's education in the nineteenth century. It is also the case, I wish to argue, for many women today who reject stereotypical career choices but then find themselves competing with men in a world where the rules have been made by men to fit in with the ways in which men are expected to behave.

A piece of research carried out some years ago, into people's views of what are the characteristics of a mentally healthy, mature, socially competent (a) adult, (b) man and (c) woman is very revealing (Broverman et al, 1970). The characteristics of a normal adult and a normal man match very closely, while those of a normal woman are quite different. It is not possible (according to these profiles, reflecting views held by both men and women) to be, at the same time, a normal woman and a normal person! This

> *... places women in the conflictual position of having to*
> *decide whether to exhibit those positive characteristics*
> *considered desirable for men and adults, and thus have their*
> *"femininity" questioned, that is, be deviant in terms of*
> *being a woman; or to behave in the prescribed feminine*
> *manner, accept second-class adult status, and possibly live*
> *a lie to boot. (p. 6)*

So, a woman working in a male-dominated and male-defined sphere, finds herself continually faced with having to choose between acting in ways which are appropriate to her as a woman, and appropriate to her as, say, an engineer. For men in these jobs there is no such conflict, whereas for women, the conflict is an inevitable part of the job and indeed of being a mature and responsible adult in a sex-stereotyped world.

The combined effect of coercive inducements and double conformity is to increase enormously the obstacles which women have to overcome when they try to make their way in male-dominated and defined areas of study and work. Competence and confidence in mathematics plays a part in many of these, and girls who opt out of mathematics, science and technology at school, because they do not wish to enter these fields, are responding to very strong influences. They can hardly be said to be making choices based only on talent, interest or inclination. Although not usually expressed in this way, many girls see the choice as between living their lives under the stress of double conformity, and being continually in a 'conflictual position' or, alternatively, gaining the rewards for conventionally feminine choices and behaviour. These rewards may well be short-term and short-lived, but life beyond age 25 is not salient in the eyes of most girls.

When I consider the gender and mathematics issue with the aid of these explanatory constructs, I find myself ceasing to be puzzled by girls' underachievement in mathematics, but rather astonished that girls and women achieve as much as they do!

If changes are to be brought about, the loss of female mathematical talent abated, and greater equality of opportunity and genuine freedom of choice opened up, then we have to look both at and beyond school practice in ways which take account of these deep-rooted forces. We have to work simultaneously on a number of fronts. One of these is to change the climate within mathematics classrooms so that ways of working which girls find comfortable are welcomed. An example of this would be to develop classroom practices and types of classroom organisation which discourage competitive behaviour, (where the search for the right answer is dominant), and instead encourage cooperative, collaborative and exploratory behaviour where each person's contribution - as an individual or as a member of a group - is valued. I find hopeful in this respect the directions in which GCSE mathematics, properly applied in the classroom, could take us. Another is to look carefully at the content of the mathematics curriculum, and ensure that this reflects a broad range of human concerns, rather than being narrowly focussed on traditionally male concerns only. These sorts of changes serve to reduce the level of conflict for girls. We have rightly, and at long last, gone beyond the days when it was believed that improving girls' participation in mathematics required that *girls* were to be changed!

In Holland, when an alternative mathematics curriculum was introduced (Math A), with a higher 'social' content and broader based applications which are more obviously and immediately relevant to pupils, the proportion of girls studying mathematics to age 18 greatly increased. (Isaacson et al, 1986). The Dutch experience suggests that changing the content of the curriculum alone, can make a significant difference - how much improvement might we see if a number of the important variables were changed at once! The 'bubble' diagram, (figure 1), offers a schematic view of the complex of interrelating variables I believe to be most salient in reaching an understanding of the influences on mathematics learning.

However, we also have to look at the genderisation of mathematics itself, hinted at above when I referred to mathematics as a 'hard' subject. Some of the most interesting work contributed to the gender and mathematics debate recently has addressed this question. Stephen Brown (1986), for example, discusses in a very persuasive way the need to humanise mathematics, not just, or even primarily for the sake of female students, but because he believes that mathematics, and ourselves as students of the discipline, would be enriched thereby. Brown bases his case to some extent on the work of the feminist psychologist, Carol Gilligan.

In an insightful and influential book Gilligan (1982) dissects Kohlberg's theory of a hierarchy of moral development and argues instead that women and men have distinctively different ways of dealing with moral problems. Women's approach is generally context-bound and human-related, while men employ a more strictly logical and impersonal mode of decision-making. Gilligan's description of the responses of two 11 year old children to a moral dilemma (should Heinz, who cannot afford to buy the drug needed to save his wife's life, steal it?) is illustrative of this theme. Jake accepts the dilemma as given, and offers an immediate 'solution' based on a simple equation "For one thing a human life is worth more than money..." (Gilligan, 1982, p. 26) while Amy sees the complexity of the problem "he

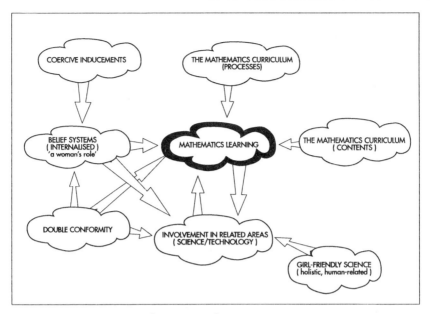

Figure 1 :   Influences on Mathematics Learning

really shouldn't steal the drug - but his wife shouldn't die either" (Gilligan, 1982, p. 28). Amy goes on to point out that if Heinz stole the drug he might have to go to jail and then would not be available to get more of the drug if his wife should need it. She seeks a solution which rejects both extreme alternatives and wishes to mediate this through discussion: "they should really just talk it out and find some other way to make the money".

In a recent paper, Leone Burton (1987), develops some of these themes and argues that

> *... the discipline itself, and the style through which it is encountered, is rendered masculine by the misguided stress laid on completeness, certainty and absolutism. From these beliefs, there develops a formal, abstract and non-applicable presentation of mathematics which justifiably alienates a majority of learners and, in particular, women. ... Our pedagogical practices in teaching mathematics deny the influence of the individual or the social context, and present young people with a pretend world of certainty, exactitude and objectivity. This pretend world is associated with power and control and, given our social history, that is perceived as male by association!* (p.5)

Work on gender and mathematics must go on not only because of arguments derived from justice (women should not be discriminated against) or because of arguments derived from need (we cannot afford to neglect so much potential talent), important though both these are, but also because through debate on this issue we begin to learn much which is of significance - both to the discipline of mathematics and to its pedagogy. We come closer to understanding the factors which influence mathematics learning in general, not just in females, and we find ourselves at the sharp edge of work which aims to develop our understanding of the nature of mathematics and the possible future shape of its pedagogy.

NOTES

1.   IOWME (International Organisation of Women and Mathematics Education) holds regular international meetings and publishes a biennial newsletter. In 1986 in London, a number of 'state-of- the-art' reports from around the world were presented, as well as information gleaned from the SIMS (Second International Mathematics Survey) data, all pointing to a continuing imbalance between female and male achievement in mathematics. In March 1987, the Dutch government funded a conference organised by Vrouen & Wiskunde on the theme Vrou is Kundig (women are capable). In the U.K., the publication by the Royal Society of a report on Girls and Mathematics (1986) is a measure of the public recognition now given to this issue.

2.   The author spent a term in an Infants School in 1986, observing gender differences in children's play, style of dress, choice of companions and activities. She also explored with 6 and 7 year old children their views on a number of questions through in-depth individual interviews.

REFERENCES

Broverman, I, Broverman, D,
Clarkson, F, Rosenkrantz, P & Vogel, S
        (1970)   'Sex-role stereotypes and clinical judgments of mental health' in *Journal of Consulting and Clinical Psychology* **Vol 34 No 1** pp 1 - 7.

Brown, S.I.   (1986)   'The Logic of Problem Generation: from Morality and Solving to De-Posing and Rebellion' in Burton, L (ed) *Girls Into Maths Can Go* ,Holt.

Burton, L (ed)   (1986)   *Girls Into Maths Can Go* ,Holt.

Burton, L   (1987)   'Women and Mathematics: Is There an Intersection?' in *IOWME Newsletter* **Vol 3 No 1**, pp 4 - 7.

Delamont, S.   (1978)   'The Contradictions in Ladies' Education' in Delamont S. and Duffin L. (eds) *The Nineteenth Century Woman: Her Cultural and Physical World*, Croom Helm.

Easlea, B,   (1981)   *Science and Sexual Oppression*, Weidenfeld and Nicholson.

Easlea, B.   (1983)   *Fathering the Unthinkable*, Pluto Press.

Gilligan C   `(1982)   *In a different voice*, Harvard University Press.

Isaacson, Z   (1982)   'Gender and Mathematics in England and Wales: A Review' in *An International Review of Gender and Mathematics*, ERIC

Isaacson, Z.   (1986)   'Freedom and girls' education: a philosophical discussion with particular reference to mathematics' in *Girls Into Maths Can Go* , Burton, L (ed), Holt.

Isaacson, Z,
Rogers P &
Dekker T   (1986)   'Report on IOWME discussion group at PME 10' in
                            *IOWME Newsletter **Vol 2 No 2** pp 3 - 6*

The Royal
Society   (1986)   *Girls and Mathematics* A report by the Joint Mathemati-
                            cal Education Committee of the Royal Society and the
                            Institute of Mathematics and its Applications, The
                            Royal Society.

Sharma, S &
Meighan, R   (1980)   'Schooling and sex roles: the case of GCE 'O' level
                            mathematics', *British Journal of Sociology of Education*,
                            1 (2).

Part 4

POLITICAL ACTION THROUGH MATHEMATICS EDUCATION

INTRODUCTION

This final section includes three papers in which aspects of a political nature are identified in relation to mathematics education.

In the first of these, Stephen Lerman examines the notions of values, power and control through mathematics education and develops the case for viewing the teaching of mathematics from a problem-solving perspective as a revolutionary activity in the true political sense. In doing so, he draws on the work of people such as Freire in connection with the role education has to play in the oppression and freedom of peoples, Bloor who links mathematics with the social construction of realities and Brown who develops the idea of problem-posing as a basis for the mathematics curriculum. Lerman concludes by identifying three potential revolutionary changes that arise from the adoption of such perspectives within the teaching and learning of mathematics.

John Abraham and Neil Bibby explore the potential for establishing a 'Mathematics and Society' curriculum within a formal education system. Current debates about the purpose of mathematics education are examined including an historical perspective of the place of mathematics within the curriculum, the development of 'ethnomathematics' and mathematics seen as a social institution. They develop the notion of a Public Educator Mathematics Curriculum for which they offer a model and go on to identify and discuss issues arising from the adoption of such a model including matters relating to political ideology, teacher education and students.

Jeff Evans examines work related to the mathematical needs of adults and develops the idea of 'community research' knowledge and skills characterised in terms of applied statistics. The notion is developed of a non-professional who has an expertise and competence which they communicate to their community. Evans draws from anthropological studies to identify distinctions with respect to different levels of dimensions required within such 'folk maths' and specifies relevant problem-solving skills. The idea of 'barefoot statisticians' emerges and appropriate content for their mathematical education is suggested with examples from work currently carried out in the U.K.

LEARNING MATHEMATICS AS A REVOLUTIONARY ACTIVITY

Stephen Lerman

".. the problem-posing educator constantly re-forms his reflections in the reflection of the students. The students - no longer docile listeners - are now critical co-investigators in dialogue with the teacher. The teacher presents the material to the students for their consideration, and re-considers his earlier considerations as the students express their own."

It is recognised in mathematics education that mathematics questions can be political, whether in an overt way or otherwise, that we should be conscious of their perhaps hidden messages and that indeed we can use mathematics work for such things as anti-racist education. It is a further and significant step to consider the relevance for mathematics education of the assertion:

"as an act of influence, education is .. an inherently
political act." [Apple 1979, p. 102]

In this paper I propose to examine in some depth the ideas of values, power and control through mathematics education, and to put forward a strong case for viewing mathematics teaching from a problem-posing position as revolutionary. In this light it is interesting to note that the extract that heads this paper is not from a mathematics education text, but is from Freire's 'Pedagogy of the Oppressed' [Freire 1972 p. 68]. This book was written as reflections on his literacy work with the oppressed people of Brazil and elsewhere in South America. The ideas he expresses, however, can also be seen to be relevant and revolutionary in our own situation. The language in which Freire's ideas are expressed above are very similar to the ones sometimes used when discussing problem-solving in mathematics teaching, but with the emphasis here on problem-posing. This too is reflective of some recent work in mathematics education, which will be discussed below. The paper will begin by drawing out of the literature of sociology of education some of the strong theses relating to emancipatory education, followed by an examination of some recent relevant work in our own field. Despite some of these texts having been around the educational world for some time, they are not well known to most of us in mathematics education [however, see Frankenstein 1987], perhaps out of a resistance to the idea that mathematics has anything to do with values or emancipation.

KNOWLEDGE AND POWER

Freire describes the traditional view of education as the 'banking' concept, whereby pupils are seen as initially empty depositories, and the role of the teacher is to make the deposits. Thus the actions available to pupils are storing, filing, retrieving etc. In this way, though, pupils are cut off from creativity, transformation, action and hence knowledge.

> "For apart from inquiry, apart from the praxis, men cannot be truly human. Knowledge emerges only through invention and re-invention, through the restless, impatient, continuing, hopeful inquiry men pursue in the world, with the world, and with each other."
> [Freire 1972 p. 58]

The alternative view of education, Freire describes as the 'problem-posing' concept. By this view, knowledge is seen as coming about through the interaction of the individual with the world. 'Problem-posing' education responds to the essential features of the conscious person, intentionality and meta-cognition.

Freire's discussion of opposing concepts of education is integrally tied with oppression and freedom. He shows firstly that the 'banking' concept is an illusion, in that students do not realise that they are educating the teacher at the same time as educating themselves. They are fooled into believing in the role provided for them by the 'banking' teacher. Secondly, he shows that this approach is necessary to the oppression of the people. They must not be allowed to come to see that the transformation of their lives is within their possibility. Just as the knowledge that is deposited with them is merely for them to file, so too is the structure of society, and their function within it. Thus the teacher, inadvertently, represents the oppressor, and reinforces the self-image of the oppressed. Thirdly, he shows that anyone committed to the liberation of the oppressed must reject entirely the 'banking' concept. He writes:

> "Liberation is a praxis: the action and reflection of men upon their world in order to transform it. Those truly committed to the cause of liberation can accept neither the mechanistic concept of consciousness as an empty vessel to be filled, nor use the banking methods of domination (propaganda, slogans-deposits) in the name of liberation. Those truly committed to liberation must reject the banking concept in its entirety, adopting instead a concept of men as conscious beings, and consciousness as consciousness intent upon the world." [Freire 1972 p. 66]

In attempting to deny the possibility of people recognising the power to transform their lives that they possess, teachers maintain the passive acquiescence of the oppressed in society, and to change this it is not sufficient to support the people with slogans: one must reveal the fundamentally revolutionary nature of the alternative, and become actively involved in it.

In an analysis of the influence of ideology in the curriculum, Apple points to similar alternatives, and also to the significance of the status of the teacher.

> "Students in most schools and in urban centers are presented with a view that serves to legitimate the existing social order since change, conflict, and men and women as creators as well as receivers of values and institutions are systematically neglected... these meaning structures are obligatory. Students receive them from persons who are 'significant others' in their lives, through their teachers, other role models in books and elsewhere." [Apple 1979 p. 102]

Apple and others [e.g. Harris 1979], indicate that one of the characteristics of ideologies is that the view of the world and the meanings given, are self-justifying. Thus, power and knowledge are linked. The only way they see of breaking out of this cycle, is through the combination of critical theoretical analysis and understanding, reflection and action. This is praxis. It is not clear however, that this Marxist analysis takes one out of the cycle, or merely into another competing ideology. If Apple maintains that the 'facts' of the world are seen through the lenses of particular theories, it is difficult to see how one removes oneself from this relativism completely, through praxis. Nevertheless, through critical analysis and reflection one can compare competing world views for their abilities to generate powerful and testable theories, and possibilities for action. Relativism need not be seen as leading to the impossibility of the notion of progress in knowledge, or the incommensurability of rival theories. On the contrary, absolutism may be seen as robbing the search for knowledge of its creativity, and of its unlimited potential. [For a more full discussion of this, see Lerman 1986].

Freire's 'problem-posing' concept of education places knowledge and power firmly in the hands of people, as against the 'banking' concept which places knowledge and power with the elite. It is in the interests of the elite in the latter concept to perpetuate this view, which brings with it the oppression of the majority for the benefit of that elite, whereas in the former the interests of all, equally, are served.

In his work in the sociology of knowledge, and in particular the sociology of mathematical knowledge, Bloor analyses the way that we use our conceptions of reality for social control. He writes:

> "We should start with the idea that in our social interactions we are always trying to put pressure on our fellows or evade that pressure. The crucial point is that in order to apply pressure more effectively we try to make reality our ally. We construe reality in such a way that it justifies or legitimates our course of action." [Bloor 1979 p. 13]

The ability to appeal to a higher authority for certainty is, in the traditional mode, a necessary tool for social control. Deviant behaviour is then clearly identified and can be excluded.

An interesting analogy has been drawn by Bloor [1978] between monster-barring techniques in mathematical development, as outlined so clearly in Lakatos's book "Proofs and Refutations" [Lakatos 1977], and the exclusion of animals which do not fit a specific categorisation in Jewish Dietary Laws, described by Mary Douglas in her book "Natural Symbols: Explorations in Cosmology" [Douglas 1973]. Bloor writes:

> "These books have a common theme: they deal with the way man responds to things which do not fit into the boxes and boundaries of accepted ways of thinking: they are about anomalies to publicly-accepted schemes of classification. Whether it be a counter-example to a proof; an animal which does not fit into the local taxonomy; or a deviant who violates the current norms, the same range of reactions is generated....
>
> The crucial point is that Mary Douglas has an explanation of why there are different responses to things which break the orderly boundaries of our thinking: these responses are characteristic of different social structures. Her theory spells out why this will be so, and describes some of the mechanisms linking the social and the cognitive. This means we should be able to predict the social circumstances which lie behind the different responses which mathematicians make to the troubles in their proofs." [Bloor 1978 p. 245]

Constructions of reality, ideologies, world views, are not merely alternative theories with metaphysical implications only, to be discussed, compared, refuted or supported in the ivory towers of philosophy. Knowledge and Power are inseparably linked, and knowledge is used as and for power, the domination of one group over another, the oppression of people, the legitimation of that oppression and the rationalization of values.

EMANCIPATORY MATHEMATICS EDUCATION

Recent literature in mathematics education has indicated a growing awareness of the part that mathematics education has to play in the perpetuation of power.

Gerdes [1985] has pointed out, from his experience in Mozambique, that mathematics is not neutral, and has to involve itself with issues of war or peace, liberation or oppression, the suppression of indigenous culture, or its fostering. Some critics have dismissed such work as romanticising native culture, but to do so is to fundamentally underestimate and misunderstand the significance. As Gerdes puts it:

> "A problemising reality approach as starting point is in itself already a confidence-creating activity. Problemising reality, reinforced by cultural, social and individual-collective confidence-stimulating ac-

> tivities will contribute substantially to an emancipa-
> tory mathematics education, to enable everyone and
> every people to understand, develop and use math-
> ematics as an important tool in the process of under-
> standing reality, the reality of nature and of society, an
> important tool to transform reality in the service of an
> ever more human world." [p. 20]

Mellin-Olsen [1985] discusses similar issues in relation to the production and reproduction of culture, and the relationship of knowledge to this process.

In mathematics education, Stephen I. Brown has been particularly influential in focusing our attention on problem-posing as distinct from problem-solving. This latter retains the emphasis on the solution, whereas the real liberation is to be found in freeing oneself from the limitations of the problem as set, and to re-pose it. He writes:

> "One can start with a definition, a theorem, a concrete
> material, data or any other phenomenon and instead
> of accepting it as the given to be explored, one can
> challenge it and in the act create a new 'it'.
> ... such activity has a built in kind of irony, for it is in
> the act of 'rebellion' that one comes to better under-
> stand the 'thing' against which one rebels. In that
> sense challenging 'the given' as a strategy for problem-
> generating has the potential to be viewed as a less
> radical departure from standard curriculum than one
> might otherwise believe." [Brown 1984 p. 19]

This last idea refers to the possibility of taking standard content and using it in the different way that he describes in that paper. He gives an illustration of re-posing an investigation of the Fibonacci sequence. In my view, however, the implications of the ideas raised by Brown, in the context described in this present paper, make the departure very radical indeed.

Cobb [1986] picks up the same theme, and deals in particular with the implications of the constructivist view of learning, in relation to the context of problem-
generation. He appears to recognise the fundamentally revolutionary character of
problem-posing. He writes:

> "Self-generated mathematics is essentially individual-
> istic. It is constructed either by a single child or a small
> group of children as they attempt to achieve particular
> goals. It is, in a sense, anarchistic mathematics. In
> contrast, academic mathematics embodies solutions
> to problems that arose in the history of the culture.
> Consequently, the young child has to learn to play the
> academic mathematics game when he or she is intro-
> duced to standard formalisms, typically in first grade.
> Unless the child intuitively realizes that standard
> formalisms are an agreed-upon means of expressing

and communicating mathematical thought they can
only be construed as arbitrary dictates of an authority.
Academic mathematics is then totalitarian mathematics." [p. 7]

Bishop [1985] refers to what he calls the "necessary power imbalance implicit in the
teaching/learning relationship", and suggests that the notions of imposition and
negotiation can be seen as opposite ends of a continuum of teacher behaviour. This
continuum is a useful construct in thinking about the kind of power imbalance that
exists in classrooms.

A SYNTHESIS

Whilst it is generally recognised that education is related to values, it is probably felt
that we in mathematics education are relatively free from such concerns. There are
some aspects of mathematics teaching that are clearly close to issues of values, by
any interpretation of mathematics education. For example, we may have to
consider what mathematics we teach to pupils of different abilities, and what
might be meant by attempting to ensure numeracy in all school-leavers, and their
ability to cope with any mathematical situations in adult life. However, if we retain
our view of mathematics itself as value-free, and of mathematical knowledge as of
a different kind to any other (because of its hold on logic, deduction, and certainty),
then we are likely to remain aloof from discussions of values in education in
general, and also perhaps merely to skim the surface of the consideration of the
possible function of mathematical education, as in the instance above. There are,
however, very good reasons for rejecting absolutism in mathematics as a research
programme. I have outlined some of these elsewhere [Lerman 1983].

Mathematics de-reified, seen as a social invention, its truths, notions of proof etc.,
relative to time and place, becomes a subject of study for sociologists. It has always
been accepted that mathematicians are material for sociological study, and so too
is 'wrong' mathematics. Relativism leads to an opening up of all of mathematics
and the history of mathematics to equal treatment. There are, for instance, socio-
historical studies of the development of Projective Geometry against the background of British scientific values and beliefs in the 19th century [Richards 1986],
the influence of Hamilton's metaphysical beliefs on his algebra work [Bloor 1979],
and of the variations in the notions of proof through history [Grabiner 1974]. 'True'
or 'right' mathematics is not seen as coming about through the inexorable logic of
certainty, the revelation of timeless truth. Prejudices, jealousies, patronage, rivalry,
metaphysical and ethical beliefs are all motivating factors for the development of
mathematics, just as any other area of knowledge.

In our present discussion, the issue of what values are presented to pupils
through different images of mathematics, what relationships are developed between pupils and teacher and indeed between people and society, through mathematics education, are subjects for analysis, evaluation and action, given this
paradigm of mathematical knowledge. In terms of evaluating the potential of a
research programme, the openness of this view of mathematical knowledge to
socio-historic, and political analysis and action is further strong support for this
proramme.

To use the terms of Freire and Brown, problem-posing and problem- generation in mathematics education are potentially revolutionary developments. The body of mathematical knowledge takes a subsidiary place in the mathematics curriculum. Its relationship to the learning of mathematics becomes that of a library of accumulated experience, and just as any library is useless to someone who cannot read, so too this library is useless unless people have access to it. When a problem is generated which reveals the need for some of this knowledge, be it multiplying decimals, standard index form, complex numbers, or catastrophe theory, if the individual recognises that such help is needed, and is available, the context and relevance and meaning of mathematical knowledge is established. Mathematical knowledge grows through formal and informal testing, that is, axiomatics and formal deduction are essential aspects of mathematical work, in that they lead to one kind of examination of ideas and conjectures, but so too is the informal testing of the search for counter-examples, etc. These are of course processes that go on, not just at the frontiers of the growth of mathematical knowledge, but at all stages of mathematics learning.

Problem-generating, or rather in this particular context problem-solving, is usually absorbed into the school mathematics syllabus, where this happens at all, without an awareness of the implications and possible consequences of this work. It is not necessarily the case that fundamental changes will take place in the mathematics classroom just because of the introduction of "problem-solving and investigations", especially as they now appear in the specifically allocated time-slot for assessment-led coursework [see Lerman 1987]. The role of the teacher as the person who knows, as the person who will eventually reveal the answer need not, and usually does not, alter. Simply at the level of opening up the mathematics lesson to discussion between pupils, and some individual work by pupils themselves, requires an awareness by the teacher of the different nature of activity that is a consequence of problem-solving work. This is an important step in itself.

The further step that may be seen to be potentially revolutionary, is that by presenting mathematics as concerned with looking at situations, (e.g. here is the Fibonacci sequence, or here is some information about education expenditure in South Africa, or here is some information about recruitment and pay-increases of the Police, Judiciary and Armed Forces in Britain during this Government) and developing and encouraging the posing of problems to be investigated mathematically, by teacher and pupils, we are making at least three important, new and revolutionary changes:

> i   mathematics belongs to everyone, and is not the esoteric, teacher- owned totalitarian subject as usually presented,

> ii   teacher and pupils are engaged together in the learning and doing of mathematics,

iii   the world at large is seen to be accessible to analysis, criticism and transformation by everyone, and we do not have to accept the way the world is, in a resigned and powerless manner.

CONCLUSION

The intention of this paper has been to present a further dimension and perspective to the discussion of the role of problem-generation, or problem-posing, in mathematics education. I have drawn upon literature that is not often considered relevant for the teaching of mathematics, but in doing so I have attempted to further demonstrate that viewing mathematics as a de-mystified, relativistic, social construction, implies that it does not stand alone in the context of a search for knowledge, but is embedded in and influenced by many other fields of study, methodologies, types of analysis etc. Also, in doing so, I have attempted to bring perspectives of the teaching of mathematics into the wider, and I have suggested revolutionary, issues of the role of education as a whole, the use of knowledge as social control, and the relationship between knowledge and power.

REFERENCES

Apple M. J.   1979   *"Ideology and Curriculum"*
                          Routledge and Kegan Paul, London

Bishop A. J.   1985   "The Social Construction of Meaning - A Significant
                          Development For Mathematics Education?" in *For The
                          Learning of Mathematics Vol. 5 No. 1*, p. 24-28

Bloor D.   1978   "Polyhedra and the Abominations of Leviticus" in
                          British Journal for the History of Science Vol.29

Bloor D.   1979   "Did Hamilton's Metaphysics Influence his Algebra?"
                          Science Studies Unit, University of Edinburgh No.4 p.
                          245-272

Brown S. I.   1984   "The Logic of Problem Generation: From Morality and
                          Solving to De-Posing and Rebellion" in *For The Learn-
                          ing of Mathematics Vol. 4 No. 1*

Cobb P.   1986   "Contexts, Goals, Beliefs and Learning Mathematics"
                          in *For The Learning of Mathematics Vol. 6 No. 2, p. 2-9*

Douglas M.   1973   *"Natural Symbols: Explorations in Cosmology"*
                          Barrie & Jenkins

Frankenstein M.
                 1987   "Critical Mathematics Education: an application of
                          Paolo Freire's Epistemology" in Ira Shor (Ed.) *"Freire
                          for the Classroom:* A Sourcebook for Liberatory Teach-
                          ing" Heinemann, New York

Freire P.   1972   *"Pedagogy of the Oppressed"*
                          Sheed and Ward, London

Gerdes P.   1985   "Conditions and Strategies for Emancipatory Mathematics Education in Undeveloped Countries" in *For The Learning of Mathematics Vol. 5 No. 1*, p. 15-20

Grabiner J. V.   1974   "Is Mathematical Truth Time-Dependent?" in American Mathematical Monthly No. 81, p. 354-365

Harris K.   1979   *"Education and Knowledge"* Routledge and Kegan Paul, London

Lakatos I.   1977   *"Proofs and Refutations"* Cambridge University Press, Cambridge

Lerman S.   1983   "Problem-Solving or Knowledge-Centred: The Influence of Philosophy on Mathematics Teaching" in International *Journal for Mathematical Education in Science and Technology Vol. 14 No. 1*, p. 59-66

Lerman S.   1986   *"Alternative Views of the Nature of Mathematics and their Possible Influence on the Teaching of Mathematics"* unpublished PhD Dissertation, King's College (KQC) University of London

Lerman S.   1987   "Investigations - Where to Now? or Problem-posing and the Nature of Mathematics" *in Perspectives No. 33*, University of Exeter.

Mellin-Olsen S.   1985   *"Culture as a Key Theme for Mathematics Education"* in Proceedings of Conference in Bergen, Norway

Richards J. L.   1986   "Projective Geometry and Mathematical Progress in Mid-Victorian Britain" in Studies in *History and Philosophy of Science Vol. 17 No. 3*, p. 297-326

MATHEMATICS AND SOCIETY:
ETHNOMATHEMATICS AND A PUBLIC EDUCATOR CURRICULUM

John Abraham and Neil Bibby

INTRODUCTION

Our primary purpose in this paper is to generate a discussion about the potential role of a 'Mathematics and Society' curriculum in formal education. The paper is probably most relevant to formal education systems in Western industrialised countries though not exclusively so. The development of such a curriculum immediately raises questions about the sort of educational philosophies underpinning the curriculum. What might be the purpose of such a curriculum? Implicitly, a 'Mathematics and Society' curriculum suggests that the study of the relationship between mathematics and society will be involved. Furthermore, questions concerning both the nature of mathematics and the nature of society need to be addressed. Thus, a 'Mathematics and Society' curriculum needs to be informed by models/theories of how mathematics is socially produced and of the role of mathematics in society as well as an underlying educational philosophy.

Much can be learned from debates about the purpose of mathematics education, both past and present. In fact, school *science* educators were the first to grapple with the problem of how the nature of the subject (in this case science) is related to educational philosophy and *vice versa*. The nineteenth-century advocates of the 'science of common things' curriculum have much in common with the proponents of modern day ethnomathematics, for instance.

Our aim is to provide a conceptual picture of the elements required for a 'Mathematics and Society' curriculum in terms of values, philosophy of mathematics, and social theory.[1] The conceptual picture should be seen as the essential product of the paper. Though it is an abstract product, it is arrived at by considering concrete curriculum debates. Also, the product should be seen as a tool which can be applied to a variety of concrete mathematics curriculum development situations.

1. THE CURRICULUM DEBATE

In nineteenth-century Britain the growth of democracy and industry with concomitant radical changes in types of work led to a fierce debate about the purpose of education. Williams [1961] has provided a useful description of the protagonists in this debate. He refers us to three lobbies, namely, the 'industrial trainers', the 'old humanists', and the 'public educators'. An expanding and dynamic economy, on the one hand, and the development of an organised working class demanding education, on the other, framed the struggle over the purpose of education. The industrial trainers pressed for a curriculum which would serve the perceived desires of industry within the economy whilst the public educators proposed an education for democratic citizenship. By contrast, the old humanists can be seen as a conservative influence who regarded the curriculum of the *status quo* as intrinsically justified. School subjects would, thus, continue to be studied for their own sake and to be geared to the interests of an elite.

In an in-depth study Layton [1973] gives a vivid account of how the old humanist, industrial trainer, and public educator lobbies influenced the development of the British school science curriculum in the nineteenth century. In the 1850s advocates of the 'science of common things' curriculum, such as Dawes and Moseley, felt that science teaching should draw upon and reflect the children's experiences of 'everyday living'. Their teaching materials thus included problems of cottage ventilation, personal hygiene, family nutrition, manual competences, and agricultural improvement.

Dawes argued that science must be made to bear upon 'practical life' which he maintained changed according to the dominating problems of each society and the immediate concerns of particular learners. In particular, scientific knowledge was to be adapted to the needs of working-class children for whom science had previously been considered of no relevance. Thus the movement for the teaching of 'science-of- common-things' reflected, in the main, a public educator perspective.

However, it also drew on arguments closely related to those of the old humanists and industrial trainers. For example, Dawes argued that the science of common things would 'raise them{pupils} in the scale of thinking beings'.[2] This is reminiscent of the old humanist view that science for its own sake refined and elevated every human feeling. Simultaneously, industrial usefulness of scientific knowledge was stressed. Moseley had a history of interests in industrial education and according to Layton supported opportunities to apply to companies such as the East India Company for financial support for the curriculum.

As Layton explains, the 'science-of-common-things' curriculum was opposed on several grounds. Firstly, to some in politically powerful positions, the idea of giving the masses access to knowledge which the upper classes would not also have, was threatening. The fear was that it might lead to social instability and subversion of the pre-existing social order. Secondly, liberal educators argued that a curriculum selected because of its immediate utility to a particular social group (namely the working class) might lead to a 'ghetto curriculum' in which pupils were discouraged from looking beyond their own environment, thus frustrating the liberal ideal of social mobility. Thirdly, modern science and its industrial applications were thought to be best served by the application of mathematics to scientific problems, especially physics. The mathematization of science was the antithesis of the 'science-of-common-things' curriculum. Ultimately, according to Layton, these arguments were enough to lead to the demise of the 'science-of- common-things' curriculum movement.

Some one hundred and fifty years later a similar discussion is growing with respect to the mathematics curriculum. Faith in the old humanist perspective that mathematics education is a rigorous training of the general faculties of the mind has diminished, and currently an expanding forum of debate regarding the purposes of mathematics education is developing.

One of the major justifications of the old humanists for mathematics as a crucial staple in the curriculum is that it offers a rigorous training in rational thought. On this argument the study of mathematics is supposed to extend logical thinking, critical thinking and problem-solving abilities. Essentially, then, mathematics in itself, and for its own sake, engenders the educated person. Amongst educators there is now considerable scepticism about this old humanist perspective in its pure form, though institutionally it remains a powerful lobby.

Further criticism of the old humanist view of mathematics education has also come from the industrial trainers. They argue that the study of mathematics for its own sake cannot be relied upon to deliver the skilled manpower for work in the economy. Mathematics education, according to the industrial trainers lobby, should be oriented towards applying mathematics to industrial problems. (Thwaites [1961]) In contrast to the old humanists, this view shows little concern for the development of the general faculties. It is characterised by an emphasis on society's declared need for specific pragmatic and instrumentalist thinkers.

What, then, can be said about the relationship between mathematics and society as viewed through the perspectives of these three educational philosophies ? For the old humanists, mathematics has no apparent relationship with society at all. For the industrial trainers, mathematics has a narrow one-way relationship in which mathematics is required to provide the techniques and expertise demanded by particular interests created through technological change and industrialisation. Neither group sees mathematics and society as having an interactive relationship. The public educator perspective, however, takes a somewhat more interactive approach in that the kind of mathematics which is seen as appropriate for the curriculum is built on a view of society which takes account of different constituencies of interests, including the cultural interests of the learner.

2. ETHNOMATHEMATICS AND SELF-GENERATED MATHEMATICS

Recently, the importance of cultural context has formed a central theme in a number of mathematics education research projects. As with the movement for teaching science of common things, a common element of these projects is that the legitimation of learners' experiences is recognised as being of fundamental pedagogical importance. In short there is a focus on 'ethnomathematics', a term coined by D'Ambrosio [1985] to refer to 'mathematics which is practised among identifiable cultural groups, such as national-tribal societies, labour groups and so on'. (D'Ambrosio [1985] p.45)

D'Ambrosio's research programme is designed to identify cases of ethnomathematics and relate them to their historical origins and patterns of reasoning. Clearly this has implications for education since it allows the possibility of a redefinition of legitimate mathematical knowledge and practices. The vision of the ethnomathematics perspective is that the 'psychological blockage' often associated with the learning of academic mathematics might be avoided. (Gerdes [1986] p.21)

Gerdes [1986] provides one of the most striking examples of ethno-mathematics in Third World countries. His demonstration of the geometrical thinking involved in Mozambican weaving illustrates indigenous 'frozen' mathematics. Gerdes' argument is that the weavers, through their activities, already engage in complex mathematical thinking. Winter [1987] makes a similar point about the 'metaphorical' thinking of young children playing number games in informal contexts. He suggests that infants and young children already possess the conceptual understanding of mathematical notions such as infinity or probability. They lack only the agreed conventions to articulate them in academic mathematical terms. Winter maintains it is the *sophisticated format* of academic mathematics rather than the underlying concepts which children do not readily engage with.

Winter's analysis closely parallels that of Hoines [1986] who emphasises the significance of children's own language in their development of mathematical concepts. She explains how unconventional units of distance derived from the pupils' labelling (e.g.'Asmundchord') aid them in measurement. [3] Again the implication is that non-academic experiences involve mathematical thinking and that these experiences in the form of modes of thought, jargon, interests or myths can be used as liberatory tools (in a psychological sense). That is, they can be part of mathematics education but simultaneously liberate learners from the tyranny of conventions and formats in academic mathematics. They can combat the 'mathophobia' and 'psychological blockage' to which Winter and D'Ambrosio respectively refer. We suggest that many of these pedagogical perceptions contained within the ethnomathematics perspective would be of considerable significance in a 'Mathematics and Society' curriculum.

Cobb [1986] spells out some other consequences of opting for ethnomathematics (what he calls 'self-generated mathematics') in preference to academic mathematics. Self-generated mathematics, he argues, is individualistic and anarchistic. The criterion of acceptability for self-generated methods is pragmatic i.e. the methods function to enable children to attain their goals. By contrast, academic mathematics requires children 'to play the academic mathematics game when he or she is introduced to standard formalisms, typically in first grade'. (Cobb [1986] p.7) Cobb charges academic mathematics with authoritarianism and totalitarianism and says:

> The child's overall goal might then become to satisfy
> the demands of the authority rather than to learn
> academic mathematics *per se*. (Cobb [1986] p.7)

Given the similarities between ethnomathematics and the 'science-of- common-things', it would not be surprising if criticisms of the latter applied to ethnomathematics. In effect, might solely giving legitimation to alternative (or self-generated) mathematical thinking in 'everyday life' lead to a 'ghettoising' of the curriculum ? In the case of the 'science-of-common-things' curriculum, liberal educators voiced this concern because they wanted to maximise social mobility. Ethnomathematics has already faced analogous difficulties. For example, Mellin-Olsen [1986] comments on how the Norwegian Social Democratic government resisted ethnomathematics in the curriculum on the grounds that it contravened the principle of equality of opportunity in the form of equal curricular content for all.

Our own reservations about ethnomathematics are rather different. However, they are also related to the problem of a 'ghettoising' curriculum. We want to consider this problem under the two related themes of ethnocentricity and critical intervention:

(i) ethnocentricity.

Unlike Gerdes, we do not want to define mathematics *as* ethnomathematics but neither do we wish to define it as academic mathematics. Mathematics is more than either of these. Mathematics is produced not only through 'everyday experiences' untouched by academic influence, but also through the organised activity of particular social groups whose mathematical problems arise because of an historical conjuncture between the groups' structural role in society and the predominant mathematical paradigms of the time. We shall call this socially organised mathematical activity *the social institution of mathematics*. Indeed it is precisely due to an appreciation of the importance of the social institution of mathematics that the industrial trainers have lobbied and continue to lobby for an industrially-orientated mathematics curriculum. Mathematics education should involve some individual and group generation of mathematical problems - this is the great insight of ethnomathematics. But we do not believe that this is sufficient for a mathematics education: in addition we wish to include that part of the public educator perspective which emphasises *democratic citizenship* rather than the liberal promises of *social mobility* or *equality of opportunity*. In our view democratic citizenship with respect to mathematics means not only having the skills to generate one's own mathematical problems but also having some understanding of how and why other pervasive mathematical problems are generated and maintained along with their most important consequences for democracy and citizenship. In short, having some understanding of the social institution of mathematics.

(ii) critical intervention.

We presume that proponents of ethnomathematics would argue for the development of mathematics as a cultural resource i.e. something which a group or subculture/culture can use as readily as speakers use their own language. However, this requires more than

a mathematics curriculum which legitimises everyday encounters with mathematics. Such legitimation
is necessary, but we argue not sufficient, for the development of mathematics as a cultural resource. To
achieve this objective, surely practitioners must be
empowered to either use, or reject the usage of, mathematical techniques by reference to their cultural value
system ? This implies that practitioners possess skills
which enable them to make value-judgements. It seems
to us, therefore, that a mathematics education which
seeks to develop mathematics as a cultural resource
should not only relate to learners' experiences but also
contain a critical dimension orientated towards making judgements about experiences on the basis of an
understanding of how context influences those experiences. This might involve, for example, students
critically comparing their non-formal types of mathematical thinking with the 'official' version of mathematics presented to them.

To clarify these arguments it is necessary to consider in more detail how critical
intervention and the social institution of mathematics relate to mathematics education.

3. THE SOCIAL INSTITUTION OF MATHEMATICS

Sociologists, historians and others have already studied many aspects of the social
institution of mathematics (e.g. Mackenzie [1981], Mehrtens *et al* [1981], Grattan-
Guinness [1981], and the Government Statisticians' Collective [1979]. For our
general and simplified view of the social institution of mathematics (largely
informed by the aforementioned works and others) see Fig.1. Sociologists of
education such as Cooper [1985] and mathematics educators such as Ernest [1987]
have also studied some of the institutional relationships between mathematicians,
mathematics educators and the curriculum. Nevertheless all these works and the
understandings gained from them have remained the province of history, sociology, educational studies or some other discipline.

What we are suggesting is that *mathematics* cannot be completely understood
without some understanding of the social institution of mathematics. This means
having an understanding of the human actions and commitments that give rise to
major developments in mathematics. It means having an understanding of the role
mathematics plays in structuring our experiences and judgements. This point is a
crucial complement to the ethnomathematics perspective because it addresses the
question of how context affects and structures experience. The ethnomathematics
perspective leaves this question unasked since 'everyday experiences' are taken as
the starting point. We believe that this is not sufficient because mathematics which
we do not directly experience can still have important consequences for our lives
and from the public educator perspective especially those aspects of our lives which

depend on democratic citizenship. We propose that the study of the social institution of mathematics becomes part of mathematics education.

After all, mathematicians experience this social institution, mathematics educators are products of it, mathematics teachers are products of it, and so on. Hence a 'Mathematics and Society' curriculum should bring the study of the social institution of mathematics *into the curriculum* as well as embrace the insights from ethnomathematics.

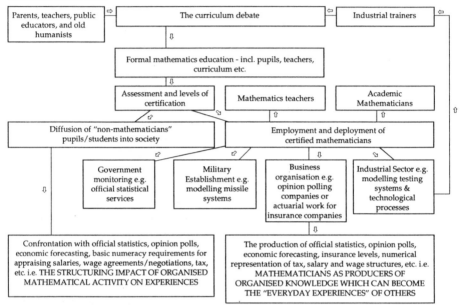

Figure 1 : **The Social Institution of Mathematics**

4. CRITICAL THINKING AND CONSCIENTIZATION

According to McPeck [1981] most people interested in educational issues tend to see the ability to think critically as a desirable human trait. The old humanists supported this view with respect to mathematics education except that they regarded it as an automatic appendage to thinking-through academic mathematical problems. This old humanist perspective is concisely countered by McPeck [1981, p.21] when he says that 'the requirements for assessing a problem critically are epistemological, not logical, in character'. It follows that expert manipulation of logical relations within the paradigms of academic mathematics gives no guarantee of, or even cognitive basis for, critical thinking.

Wellington and Wellington [1960] go futher by asserting that critical thinking arises out of problem-generating and problem-solving activity. The definition of the problem is not the teacher's but rather it is considered to arise from the anxieties [4] of the students in areas of interest to both teacher(s) and students.

> The process of planning and sharing by teacher and
> students helps to produce coalescence, the basic method
> of critical thinking. (Wellington and Wellington [1960]
> p.85)

This model of critical thinking has some similarities to the ethnomathematics perspective but it can be contrasted with Cobb's self-generated mathematics in the important sense that critical thinking is considered to be *cooperative*, rather than *individualistic*. Problems are *shared* by individuals rather than *possessed* by them.

It is in the work of Harris [1981] for UNESCO and Freire [1972] that conceptions of critical thinking best bring together the ethnomathematics and public educator perspectives. Harris's UNESCO report lists the need for active and critical participation in the democratic process as one purpose of mathematics education. Similarly, Freire [1972] is concerned to promote 'free critical citizenship' through the educational process of conscientization. [5] But what do we mean by the Freirian concept 'conscientization'? Mellin-Olsen [1986] gives the concise interpretation that conscientization is the process by which people are made aware of their culture.

To be more precise, conscientization can be understood as the process by which people become aware of how their experiences are structured and conditioned. This awareness enables people to make critical choices about actions. As Freire puts it:

> Conscientization is viable only because men's [6] consciousness, although conditioned, can recognize that it is conditioned. This 'critical' dimension of consciousness accounts for the goals men assign to their transforming acts upon the world. (Freire [1985] pp.69-70)

For a 'Mathematics and Society' curriculum conscientization is the crucial process by which the relationships between mathematics and society (especially the social institution of mathematics) are related to the personal development/situation of the pupils or students. The process involves the learner in a number of stages. Firstly, engagement with some form of organised mathematical activity. For pupils/students of mathematics this is immediate.[7] Secondly, objectification of some mathematical problem, i.e. the distancing of oneself from the problem so that it is seen clearly as the object of study. Thirdly, critical reflection upon the purpose and consequence of studying this problem in relation to wider values.

As informed by the ethnomathematics and public educator perspectives, the aim of the 'Mathematics and Society' curriculum would be to continually link knowledge of relationships between mathematics and society to student personal and collective development. For example, the curriculum would encourage the explanation of how controversial issues can be discussed and sometimes resolved with the aim of enabling students/pupils to develop frameworks, concepts and approaches to guide action. Teachers would try to go beyond a pedagogy which merely presented a range of alternative experiences and opinions because this would provide no insight into how to arrive at conclusions critically or make choices. In this respect teachers would try to generate an awareness of the social responsibilities of mathematicians as opposed to 'an isolationist view which divorces mathematics from its social and political context'. (Ernest [1986] p.17)

We are now in a position to consider the *conceptual* scheme underpinning the 'Mathematics and Society' curriculum being proposed. Mathematics exists in society as a socially organized activity (the social institution of mathematics) and as *ad hoc* experiences. The former structures our mathematical experiences in well-defined ways which have been established through the historical development of certain forms of organization. The latter may be used as everyday resources to generate ethnomathematics. But structured mathematical experiences can also impact upon the generation of ethnomathematics since they too are usually part of a subculture's mathematical resources. Finally, if and when ethnomathematics becomes established it can then become a structuring agent of the mathematical experiences of those outside the subculture which generated it.

Fig. 2 is a diagrammatical representation of the role of conscientization in such a curriculum. Essentially our approach differs from the ethnomathematics perspective in the sense that we wish to see a mathematics education which, in part, aims to enable students/pupils to understand how knowledge is established (including, and sometimes especially, in those spheres of social activity of which they have no immediate experience) and critically relate this understanding to their own experiences.

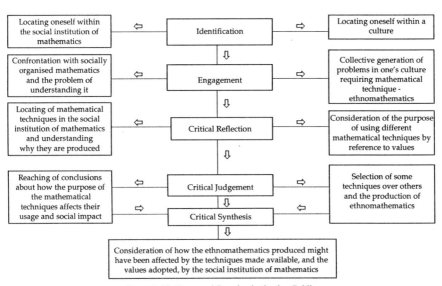

Figure 2 : **The Process of Conscientisation in a Public Educator Mathematics Curriculum**

5. EXAMPLES OF 'MATHEMATICS AND SOCIETY' TYPE CURRICULA
The idea of relating mathematics to society is not new. One of the most impressive and exciting curriculum developments in this area is the international 'Mathematics In Society Project'(MISP). It is being developed in the UK, USA and Australia mainly for the secondary school level. This project is based around eight themes, namely (a) political development, (b) economic development, (c) the natural world, (d) science and technology, (e) art, (f) sport and recreation, (g) structure of modern society and (h) what people do. The project started in 1980 and is currently being introduced into some secondary schools.

Rogerson [1986] outlines much of the thinking behind the MISP project. He argues that 'mathematics as used in society' (what he labels M2) is not the same as 'school mathematics' (what he labels M1). M1, says Rogerson, takes the form of standard syllabus lists so that the 'living body' of mathematics as evidenced by its usage in society (i.e. M2) is reduced to a skeleton. Consequently, as Rogerson points out, school mathematics is more analagous to a dictionary than to a language. Just as we have argued for the conceptualisation of a 'Mathematics and Society' curriculum, Rogerson maintains that in order to see how M2 is different from M1 it is necessary to ask: 'What is mathematics ?' One of the important insights of MISP is that it tries to answer this question by using an 'extensive definition', i.e. one which is made up from the uses of mathematics in society.[8]

Inchley [1985] emphasises how social change (e.g. the introduction of calculators and micro-computers or changes in currency) provides a major rationale for studying mathematics in society. Simultaneously, she argues for social change to be used as a resource for MISP curriculum developments. Similarly, Romberg [1985] suggests that the history of mathematics should be a resource which provides students with a 'story' of mathematics. In general, MISP protagonists are keen to relate mathematics to the 'real world' even though the nature of the 'real world' is rarely debated. [9]

MISP is certainly a progressive movement within mathematics education and may be the most significant one with respect to 'Mathematics and Society' curricula to date. We share Rogerson's concerns about the 'skeletonising' of mathematics via traditional school mathematics.[10] However, there are, in our view, some problems in adopting MISP as a model for the development of 'Mathematics and Society' curricula.

In the first instance MISP is justified on the grounds that (i) it is likely to provide 'motivation' for pupils studying mathematics, thereby reducing the frequency with which it is found to be 'difficult', and (ii) mathematics is widely used and utilised in society. The problem is that neither of these justifications can be an educational rationale. Motivation *per se* cannot be an educational rationale because opposing rationales can take different stances with respect to the same kind of motivation. For example, Freirian approaches to mathematics education might oppose a curriculum designed to motivate students to practise mathematical techniques in the absence of any understanding of the purpose of the activity whilst the old humanist philosophy might equally want to encourage this kind of motivation. Merely the fact that mathematics is widely used in society cannot be an educational rationale either because several opposing philosophies can equally use this fact as support for their case e.g. public educators and industrial trainers. The fact that mathematics is widely used is merely an observation and no more. Since neither motivation *per se* nor the pervasiveness of mathematics can form an educational rationale singly or together, it is difficult to see how they provide the justification for MISP that Rogerson claims they do. This difficulty highlights the importance of trying to relocate the current debates about the purpose of mathematics education to debates about educational philosophy.

A second problem with MISP is the absence of a model of social relations. This problem arises most vividly in the MISP distinction between topics according to the two headings - 'Maths content' and 'Society content'. The MISP method for formulating questions for the curriculum seems to be to extract the 'maths content' from the topic and ask questions about the 'maths content' but in the context of the 'society content' (Rogerson [1982]). Many questions for the curriculum can be generated by this method. However, this method reveals little about *society*. The student learns little about society, and mathematics *as part of society*. This is because society is represented as fragments of, and fragmented, activities within which mathematics 'hangs around' and gets used. The result of this is two-fold:

(a)   the nature of society is apparently considered unproblematic and, therefore, some of the MISP questions carry profound sociopolitical assumptions which are not presented as open to discussion (e.g. maximization of profit).

(b)   the process of conscientization tends to be dislocated because there is no framework to guide critical judgements made on the basis of cultural values. This framework is absent because rather than the role of mathematics in society being studied in terms of its human activity *as understood within society* it is resurrected and presented within a range of apparently unrelated value-free activities. However, a refreshing exception to these criticisms is the example of modelling the energy crisis presented by Fishman [1985].

We would suggest that some of these problems could be addressed by the incorporation of a model of mathematics in society such as the 'social institution of mathematics' model. Having said this a great deal of innovative work has been completed through MISP. Some of the ideas are very encouraging and have brought forward the possibility for further work in the area.

Another less extensive example of a 'Mathematics and Society' type curriculum is the Norwegian 'Mathematics, Nature and Society' (MNS) project developed for secondary schools in the county of Hedmark in Eastern Norway (Kvammen [1986]). This project emphasises the critical selection of mathematical techniques to help in the management of environmental and social problems. In so doing it raises questions about the nature of these problems, for instance the problem of how energy resources are allocated and how they can be conserved, and of what agricultural resources are needed to sustain a certain level of production and how these should be organized to reduce long term starvation.

The MNS project appears to give more emphasis to how context affects experiences and the environment than does the MISP. There also seems to be a greater tendency to relate the purpose of the project work in the curriculum to explicit value positions e.g. uncontrolled exploitation of the environment at the expense of agricultural land is undesirable because such land is required to produce food therefore it is worthwhile to use mathematical techniques to monitor the extent of this exploitation with a view to informing future discussion and action on the issue. The one major difference between MNS and the 'Mathematics and Society' curriculum which we propose is that the pupils studying MNS are not given a framework through which to judge the impact of their mathematical competences on the wider aspects of their future lives. This is because of the absence of any engagement with the social institution of mathematics in the MNS project.

MISP and MNS involve students/pupils in project work and the assessment is project-oriented. However, it is possible to find 'Mathematics and Society' issues raised in traditional-style examination papers. In 1986 the 'Secondary Mathematics Individualised Learning Experiment' (SMILE) under the English London Regional Examining Board included a question on quantitative estimates of military spending based on figures published by the Swedish International Peace Research Institute (SIPRI) in its Certificate for Secondary Education examination. Among other things, candidates were asked to compare the average growth of military expenditure per year in the USA with that of the USSR for the period 1980-1984. They were also asked to calculate military spending per head of population in the two countries in 1980 and to complete a graph using the SIPRI table of data. At the end of the question candidates were asked to comment on their results.

This is a considerable improvement on many tables of data in mathematics questions in that the source was actually noted - implying at least that somebody *produced* the data. Indeed, in some quarters the question has been celebrated. (Anon.[1987]) However, we consider the question to illustrate a number of problems relevant to the teaching of controversial issues in a 'Mathematics and Society' curriculum. For us the main problem with the question is that the data tends to be treated as the final facts and, therefore, did not guarantee the development of a critical perspective on the data. No rival data were presented and thus the ways in which military spending is controversial were not conveyed in the question. Different groups with different institutional interests make different claims about military spending and they use different measurements and data sets to support these claims. No doubt many of the question's deficiencies are related to examination constraints.

In the absence of examination constraints we would suggest the following alternative which is based on a public educator perspective embracing the notion of conscientization. Rival sources of data could be provided and following an analysis of the differences between the data sets there might be a discussion of why these differences have arisen. This discussion might involve the development of a framework which describes the different institutional interests involved and the way these can affect the production of statistics (i.e. again, how context influences experiences). Following this students/pupils would be asked to make critical judgements about the likely reliability of the different data sets and to reach some conclusions about the controversy with a view to guiding their future actions. This process would probably involve the use of some further mathematics as well as the use of the understanding of this part of the social institution of mathematics generated by the previous discussion.

Clearly this kind of alternative is not a possibility in traditional style examination questions. As an examination question the SMILE contribution is a welcome shift from the 'unreality' of examples often found in socalled "relevant mathematics" questions (Howson [1983]).

Our view is that a concerted effort to develop a 'Mathematics and Society' curriculum of the sort we envisage can, in principle, make a crucial contribution to mathematics education and improve on some of the important developments already achieved. However, it is helpful to contrast two kinds of arguments in favour of a 'Mathematics and Society' curriculum - the 'strong case' and the 'weak case'.

The strong case holds that mathematics can only be understood when its social and historical origins are also understood. Under the strong case the mathematics of a computer progamme, for example, can only be understood if the reasons for the design and production of the programme are also understood. On the other hand, the weak case holds that mathematics cannot be completely understood unless that understanding partially involves an awareness of how mathematics is socially organised, produced and maintained throughout history and in the context of cultural influences. In the absence of such an awareness, under the weak case, the mathematics of a computer programme are understood but only partially, and possibly in a way which is socially dislocated. This kind of incomplete understanding is similar to the notion of 'semi- intransitive consciousness' which characterises a state of being in which one has only a fragmented, localised awareness of one's situation. (Frankenstein [1983], p.318). We support the weak case and this can have some implications for the policy analysis, although it does not prescribe one policy or another.

6. POLICY ISSUES FOR A PUBLIC EDUCATOR MATHEMATICS CURRICULUM

No doubt there are many who do not share our educational perspective on mathematics or a 'Mathematics and Society' curriculum. We do not assume that our arguments have created a consensus in support of our perspective. Nevertheless, leaving fundamental differences aside for the moment, we would like to consider the many practical problems of introducing such a curriculum into the formal education system.

We would suggest that at least five distinct policy problems arise in considering the development of a public educator mathematics curriculum. These are (i) political ideology (ii) subject maintenance (iii) teacher education (iv) student expectations and the examination system (v) differentiation and ability stereotyping.

(i) Political Ideology. We should not underestimate the extent to which particular political interests would attempt to undermine a curriculum which sought to critically debate the relationship between mathematics and society. Critical debate is threatening to those interest groups who wish mathematics education to *serve* their interests directly or indirectly. The examination question which we discussed earlier could hardly be accused of bringing critical debate into the examination room. At most it implied, though not explicitly, that some kind of social issue was reflected in the SIPRI data. Yet the response in the more conservative British press was definitively hostile. For example, *The Daily Mail* headline read 'Six thousand pupils take the 'propaganda' test' and asked the question: 'What has arms spending to do with a maths exam?' *The Sun*, another British newspaper, described it as 'sinister' and concluded that political propaganda, Left or Right, has no place in the classroom. The Hillgate Group, in its *Radical Manifesto*, referred to this particular examination question as 'downright propaganda' and complained that "even mathematics and music are to be given a 'peace' or 'global' emphasis" resulting in 'a gradual pollution of the whole curriculum by practices which are profoundly diseducational' (Cox *et al* [1987]). As a result of the conservative disapproval of this examination question it was decided that an examination board should vet future mathematics papers for political content (Brown [1986]).

On the basis of this response to one examination question, it is reasonable to conjecture that a widespread curriculum which sought to relate mathematics to society, including controversial political issues, would confront ideological opposition. The best suggestion that we can make to counter this is to pre-empt such ideology in the proposals and negotiations for curriculum change. There is little point in pretending that this ideological opposition does not exist and hoping for the best.

(ii) Subject Maintenance. It has been argued that subject areas are the product of social forces which, in practice, legitimate certain aspects of curriculum content and constrain others. (e.g. Fensham [1980], Goodson [1983], and Young [1976]). For example, university mathematicians and professional mathematicians, in general, may have an interest in maintaining the present discipline of mathematics. Given our adherence to the weak case for a 'Mathematics and Society' curriculum, one policy which might avoid such resistance could be to develop 'Mathematics and Society' as a separate and distinct subject. There are at least two problems with this, however. Firstly, the separate subject might be marginalised from the rest of the mathematics curriculum. Secondly, in many countries, especially Britain, the curriculum is already 'overcrowded' (O'Conner [1987]). Cramming a new subject into the present system faces obvious difficulties in this respect.

(iii) Teacher Education. The public educator curriculum we have proposed has radical implications for pedagogy. One powerful obstacle to the development and sustaining of such a pedagogy is the authoritarian 'custodial pupil control ideology' that many teachers bring to the learning situation (Denscombe [1982]). There is now considerable research indicating that, in the UK, USA and Australia, teacher training fails to equip teachers with the capacity to construct and/or sustain an alternative to the custodial control approach in the face of practical and professional pressure to be seen as a competent teacher.

Teachers are central agents in the educational process and this means that teacher education must give a much more powerful and empowering legitimation to alternative dialogical pedagogies if a public educator curriculum is to really function.

(iv) Students' Expectations and the Examination System. Even if teachers and mathematicians can be persuaded of the value of a public educator mathematics curriculum, students might not be convinced. Insofar as our public educator curriculum includes an ethnomathematics perspective, student resistance should not, in principle, be a problem since the mathematical problems are generated from their own experience. But if this curriculum is to have any impact on schools, its developers have to take account of certain pre-existing features of schools.

For example, the status of school mathematics is such that high-attaining students gain the opportunity to enter high-status universities. Under these conditions there is a tendency to promote what Holt [1969] refers to as 'producer thinker strategies'. According to Holt a 'producer' is a student who is only interested in getting right answers and who makes more or less uncritical use of formulae to get them. In this respect the comments of Reid [1984] are worth bearing in mind:

> Students, as rational consumers, are less concerned
> with knowing than with the status that comes from
> categorical membership and the future promise that
> this implies.

It is precisely this mismatch between the expectation of students as influenced by the competitive credit-collecting examination system and the educational ideals of public educators that leads to 'pupil resistance to curriculum innovation in mathematics' (Spradbery [1976]). This 'diploma disease' is particularly marked in some Third World countries (e.g. Sri Lanka) and constructing policies to counteract it is extremely difficult. (Dore [1976])

We suggest that one strategy would be to hive off entirely part of the mathematics curriculum from the examination system and have this taught and assessed by other means. The idea is that this would be the 'Mathematics and Society' part of the curriculum. This is consistent with the policy option considered in (ii) and is, again, based on the weak case for a 'Mathematics and Society' curriculum.

(v) Differentiation and Ability Stereotyping. We know from sociological research that differentiating pupils according to their 'ability' can lead to 'polarisation' effects amongst the student population (Lacey [1970], Hargreaves [1967], Ball [1981]).

Recent research also indicates that mathematics teachers, in particular, see a direct correspondence between the 'ability' with which they label students and the hierarchical structure of mathematical knowledge with which they label the subject (Ruthven [1987]). To avoid these difficulties students of 'Mathematics and Society' should not be ordered according to some external ability criteria. This, of course, necessitates other policies such as (ii) and (iv) also being implemented. Students would be encouraged to learn cooperatively and collectively without differentiation. Again this needs to be appreciated in unison with a teacher education policy which tries to combat the polarising effects of ability stereotyping and labelling by teachers.

It is not appropriate for us to propose any precise policies in this paper. Ultimately particular policies depend on specific contexts and situations which affect what is possible for various protagonists. Instead we have offered general guidelines for a public educator mathematics curriculum policy. The changes which we have identified as desirable for policy implementation are extremely challenging, some would say impossible. But irrespective of the possibilities for implementation we would suggest that our proposals provide a reference framework for the development of a 'Mathematics and Society' curriculum. Even if some diversions from what is suggested here are required for implementation, a reference point still remains by which to judge the extent of the success of that implementation. On the other hand, those who support our suggestions might argue that it is the education system which should be changed rather than these ideas about the curriculum.

ACKNOWLEDGEMENTS

This paper was first presented at the 'Research Into Social Perspectives On Mathematics Education' conference at the Institute of Education, University of London on 7th July 1987. We are very grateful to all the participants of that conference for their comments. We would also like to thank Barry Cooper and Jeff Evans for additional helpful responses.

NOTES

[1] It is implicit in this statement that we do not believe education can be value-free.

[2] Quoted from Layton [1973] p.187

[3] 'Asmundchord' is the name which the class gave to a chord made by Asmund, one of the boys in the class.

[4] It is important to distinguish between the sense in which Wellington and Wellington [1960] use the word anxiety and the neurotic sometimes associated with mathematics learning and numeracy.

[5] We are not the first to relate the Frierian notion of conscientization to mathematics education - see Abraham [1982], Frankenstein [1981], Frankenstein [1983], and Mellin-Olsen [1984]

[6] We assume that Freire's 'men' should be read as 'human' or 'people' - for us his ideas apply as much to women as men.

[7] N.B. Engagement is not the same as 'motivation'. True motivation can come, surely, only after critical reflection.

[8] Rogerson contrasts an 'extensive definition' with an 'intensive definition'. An 'intensive definition' , he explains, 'seeks to produce in a small number of words an understanding or concept of mathematics'. (Rogerson [1986] p.613)

[9] For more detailed information about MISP readers might wish to consult *Mathematics in Society: The Real Way to Apply Mathematics ?* or *1985 MISP Report.*

[10] See Abraham and Bibby, 'Human Agency: The Black Box of Mathematics in the Curriculum'. Forthcoming.

REFERENCES

Abraham, J.	[1982].	Mathematics, Critical Thinking and Conscientization. Unpublished B.Sc. Dissertation, University of Sussex.
Anon.	[1986]	Six thousand take the propaganda test. *Daily Mail* 14th June.
Anon.	[1987]	The Third R: The politics of mathematics. *Libertarian Education*, Spring, 14-15.
Ball, S.	[1981]	*Beachside Comprehensive.* Cambridge University Press, Cambridge.
Brown, P.	[1986].	Row as maths CSE examines arms spending. *Guardian* 14th June.
Cobb, P.	[1986]	Contexts,Goals,Beliefs,and Learning Mathematics. *For the Learning of Mathematics*, 6, 2, 2-9.
Cooper, B.	[1985]	*Renegotiating School Mathematics: A Study of Curriculum Change and Stability. Studies in Curriculum History* 3. Falmer Press, Barcombe.
Cox *et al*	[1986]	*Whose Schools ? A Radical Manifesto.* Hillgate Group, London.
Daily Mail	[1986]	Six thousand pupils take the "propaganda" test. 14th June.
D'Ambrosio, U.	[1986]	Ethnomathematics and its Place in the History and Pedagogy of Mathematics. *For the Learning of Mathematics* 5,1, 44-48.
Denscombe, D.	[1982]	The 'Hidden Pedagogy' and its Implications for Teacher Training. *British Journal of Sociology of Education*, 3, 3, 249- 265.
Dore, R.P.	[1976]	*The Diploma Disease.* Allen and Unwin, London.

Ernest, P.　　　[1986]　　Social and Political Values.
　　　　　　　　　　　　Mathematics Teaching, 116, 16-18.

Ernest, P.　　　[1987]　　Mathematics, Education and Society. Unpublished
　　　　　　　　　　　　paper given at the 'Research Into Social Perspectives
　　　　　　　　　　　　On Mathematics Education' conference at the Institute
　　　　　　　　　　　　of Education, University of London, 7th July 1987.

Fensham, P.J.　[1980]　　Constraint and Autonomy in Australian Secondary
　　　　　　　　　　　　Science Education.
　　　　　　　　　　　　Journal of Curriculum Studies. 12, 3, 189-206.

Fishman, J.　　[1985]　　Developing Classroom Applications Based On Socio-
　　　　　　　　　　　　economic Problems. *1985 MISP Report*, 20-26.

Frankenstein, M.[1981]　A different Third R: Radical Math.
　　　　　　　　　　　　Radical Teacher, 20, 14-18.

Frankenstein, M.[1983]　Critical Mathematics Education: An Application Of
　　　　　　　　　　　　Paulo Freire's Epistemology. *Journal of Education*, 165,
　　　　　　　　　　　　4, 315-339.

Freire, P.　　　[1972]　　*Pedagogy of the Oppressed*.
　　　　　　　　　　　　Penguin Books, Harmondsworth.

Freire, P.　　　[1985]　　*The Politics of Education*.
　　　　　　　　　　　　Macmillan, Basingstoke and London.

Gerdes, P.　　　[1986]　　Mathematics and Curriculum Development in Mo-
　　　　　　　　　　　　zambique. In: Hoines, M.J. and Mellin-Olsen, S. (eds)
　　　　　　　　　　　　Mathematics and Culture.
　　　　　　　　　　　　Caspar-Forlag, Bergen, 15-41.

Goodson, I.F.　[1983]　　*School Subjects and Curriculum Change*. Croom Helm,
　　　　　　　　　　　　Beckenham, England.

Government
Statisticians'
Collective　　　[1979]　　How Official Statistics are Produced: Views from the
　　　　　　　　　　　　Inside. In: Irvine, J. *et al* (eds) *Demystifying Social
　　　　　　　　　　　　Statistics*. Pluto Press, London, 130-151.

Gratton-
Guinness, I.   [1981]   Mathematical Physics in France, 1800-1840: Knowledge, Activity and Historiography. In: Dauben, J.W. (ed) *Mathematical Perspectives*.
Academic Press, New York, 95-135.

Hargreaves, D.H.[1967]   *Social Relations in a Secondary School.*
Routledge and Kegan Paul, London.

Harris, R.   [1981]   *Studies in Mathematics Education.* UNESCO.

Hoines, M.J.   [1986]   On the Value of Children's own Language in their Conceptual Development. In: Hoines, M.J. and Mellin-Olsen, S. (eds) *Mathematics and Culture.*
Caspar-Forlag, Bergen, 42-45.

Holt, J.   [1969]   *How Children Fail.* Penguin Books, Harmondsworth.

Howson, G.   [1983]   Motivating Maths. *Times Educational Supplement* 4th Nov. p15.

Inchley, C.   [1985]   The Changing Course. *1985 MISP Report*, 12-13.

Kvammen, P.I.   [1986]   Forest Mathematics: A Project for Secondary Schools In Areas Where The Production Is Based On Timber. In: Hoines, M.J. and Mellin-Olsen, S. (eds) *Mathematics and Culture* op. cit. 46-58.

Lacey, C.   [1970]   *Hightown Grammar.*
Manchester University Press, Manchester.

Layton, D.   [1973]   *Science for the People.* Allen and Unwin, London.

Mackenzie, D.   [1981]   *Statistics in Britain: 1865-1930.*
Edinburgh University Press, Edinburgh.

McPeck, J.E.   [1981]   *Critical Thinking and Education.*
Martin Robertson, Oxford.

Mehrtens, H.
et al (eds)   [1981]   *Social History of Nineteenth Century Mathematics.*
Birkhauser, Boston-Basel-Stuttgart.

Mellin-Olsen, S. [1984]   The Politicization of Mathematics Education. Lecture at University of London Institute of Education. Nov. 28th.

Mellin-Olsen, S. [1986]   Culture as a key theme for mathematics education. In: Hoines, M.J. and Mellin-Olsen, S. *Mathematics and Culture*. op. cit. 99-116.

O'Connor, M. [1987]   Squeezing the syllabus. *Guardian* 20th January.

Reid, W.A.   [1984]   Curricular Topics As Institutional Categories: Implications for Theory and Research in the History and Sociology of School Subjects. In: Goodson, I.F. and Ball, S.(eds) *Defining the Curriculum: Histories and Ethnographies*. Falmer Press, Barcombe, 67-75.

Rogerson, A.(ed)[1982]   *MISP: The Real Way to Apply Mathematics ?*

Rogerson, A. [1984]   The Mathematics In Society Project. *The Australian Mathematics Teacher*. 18-19.

Rogerson, A. [1986]   The Mathematics in Society Project: a new conception of mathematics. *Int. J. Math. Educ. Sci. Technol.* 17, 5, 611-616.

Romberg, T. [1985]   *1985 MISP Report*, 17-18.

Ruthven, K. [1987]   Ability Stereotyping in Mathematics. *Educational Studies in Mathematics*, forthcoming.

SMILE. [1986]   London Regional Examining Board. C.S.E. Mathematics; paper 1, 11th June 1986.

Spradbery, J. [1976]   Conservative pupils? Pupil resistance to curriculum innovation in mathematics. In: Young, M.F.D. and Whitty, G. (eds) *Explorations into the politics of school knowledge*. Nafferton Press, Driffield, England, 236-243.

Steiner, H. [1987]   Philosophical and Epistemological Aspects of Mathematics and their Interaction with Theory and Practice in Mathematics Education. *For the Learning of Mathematics* 7, 1, 7-13.

Thwaites, B. [1961]   *On Teaching Mathematics*. Pergamon, London.

Wellington, C.B. & Wellington, J. [1960]   *Teaching for Critical Thinking*. McGraw-Hill, New York-Toronto-London.

Winter, R.   [1987]   Mathophobia,Pythagoras and Roller-Skating. *Essex Papers In Education*. Vol.2.

Williams, R.   [1961]   *The Long Revolution*. Chatto, London.

Young, M.   [1976]   The Schooling of Science. In: Young, M.F.D. and Whitty, G. (eds) *Explorations into the politics of school knowledge*. Op. cit. 47-61.

MATHEMATICS FOR ADULTS — COMMUNITY RESEARCH AND "BAREFOOT STATISTICIANS"

Jeff Evans

SUMMARY

This paper considers possible maths /statistics curricula for adults. Here "adults" are understood to include "adolescents" and others who : (i) have a substantial range of social relations and everyday activities, independent of their families; (ii) have at least the opportunity for paid or voluntary work; and (iii) are conscious of having social and political interests. (Thus, in the teaching situation, we often hear the question "How is this useful to me?".)

I begin by considering adults' "mathematical needs", as investigated in surveys, observational studies, and in "task analyses". Some of the latter (e.g. Cooper(1986) in the U.K., Gerdes (1985) in the "developing world") suggest that there is a need for knowledge and skills in what might be called "community research". By this is meant producing, or retrieving, information that is relevant to (i), (ii) or especially (iii) above.

I argue that "community research" knowledge and skills might be developed by a mathematics course with an applied statistics flavour. And that as a pedagogic context or "microworld", aplied statistics would constitute an "appropriable mathematics" in the three senses set out by Papert (1980) - that is, it should be personally meaningful, empowering, and have cultural resonance (p.54).

I go on to develop the notion of a user/producer of empirical information who is not a professionally-trained statistician - much less a mathematician - but whose expertise lies in a balance of technical competence, and an ability to communicate with their colleagues or community - to be aware of the latter's information needs, to formulate an appropriate strategy for responding to those needs, and to report the results of their searches in clear terms, comprehensible to their "constituency". Drawing on parallels with "barefoot doctors" working in China and elsewhere, I suggest that these user/producers be called " - barefoot statisticians - ".

I consider what a barefoot statistician's training might be like, and what resources are available to barefoot statisticians, practising and potential.

1. NUMERATE SKILLS NEEDED BY ADULTS IN EVERYDAY LIFE, AND IN WORK

What skills, then, are needed by a "numerate" adult? Cockcroft lists among the mathematical needs of everyday life:

* the ability to read numbers and to count;
* to tell the time and to understand straightforward timetables;
* to pay for purchases and to give change;
* to weigh and measure;
* to understand simple graphs and charts; and
* to carry out any necessary calculations associated with these. (Cockcroft Committee, p. 10)

Further on, a "foundation list of mathematical topics" for a basic course at school includes the following aims (among others):

* to encourage a critical attitude to statistics presented in media;
* to appreciate basic ideas of randomness and variability, and to know meaning of probability and odds in simple cases;
* to understand the difference between various measures of average and the purpose for which each is used. (Cockcroft Committee, pp. 135 ff.)

Clearly, the Cockcroft Committee's specification of numeracy reflects a number of concepts and skills that would be considered "statistical". This is despite the fact that, of the submissions from various interested groups and individuals which they received, "surprisingly few ... made direct reference to the teaching of statistics" (Cockcroft Committee, p. 234).

There are several ways of studying what "skills" are needed by adults for their various activities: (i) **by asking** them, or their employers; (ii) by **observing** them at work; (iii) by doing a **"task analysis"** of, say, a job description.

2. SURVEYS

There appear to be few studies of the numerate skills needed by adults in everyday, non-work, situations; Brigid Sewell's (1981) study of the self-reports of adults in the U.K. was done as part of the Cockcroft Commission's investigations.

There are rather more studies of adults' needs for numerate skills in work. In London, Mary Harris (1986) has summarized the skills reportedly needed by people working in a wide range of occupations in London.

Brigid Sewell's study involved 107 adults chosen by what might be called "recruited snowball" sampling, though there was widespread reluctance to be interviewed about maths. Many of the questions have little to do with statistics, for example, whether they checked their change, or filled in a tax return, or paid the electricity bill, and how they knew they had enough money to pay for the items in their trolley at the supermarket. Many others are closely related, such as:

* whether they measured material for decorating at home (91% said yes)
* whether they weighed anything, in cooking or otherwise (86% said yes)
* whether they had access to a calculator (67% said yes) and used it (45% said yes)
* whether they looked at timetables and used maps (about 75% for both said yes)
* anything they did in their spare time that involved numbers (78% gave

examples which included calculating car running costs, planning household alterations, betting, playing card games, being involved in collective bargaining, playing management games)

Notable findings were the wide variety of methods used by individuals, "contradicting the view that there is a 'best' method in real life, even if there seems to be only one in school lessons" (A.C.A.C.E., 1982, p. 33). Also, there was apparently no correlation between the extent to which an individual used maths, and the length of their initial education, or their social (occupational) class.

The only survey of the - statistical - skills needs of adults, of which I am currently aware, is Holmes (1985). This study investigated such needs (i) from the point of view of managers, training officers, etc. and (ii) as reported by young (18-25 years) employees themselves. Again, both were what might be called "recruited snowball samples", with an apparent selection bias towards those with better educational backgrounds and with a relatively high contact with statistical ideas (p. 15). The three areas of essential background knowledge mentioned most frequently by the managers, etc., were: numeracy, literacy, and computer awareness (pp. 3-5).

The young employees were asked to indicate whether or not they used each of some 54 statistical ideas or methods; and summarising the results led to a ranking according to reported use. Notable features of the results include:

1)   The "top ten" most-used statistical skills had to do with producing data by counting or measuring, or by extracting data from larger tables, checking the accuracy of another's data, producing and interpreting tables of data and writing reports based on data for others to use.

2)   The mean was the only statistical calculation to appear in the top ten; most of the next ten were to do with graphs, tables, and the range.

3)   The use of probability and standard statistical distribution (normal, binomial, uniform) came well down in the rankings.

Such surveys as those reported above produce very interesting results, and we must hope that more work will be done in this area. Nevertheless, several formidable methodological problems need attention in the design of such studies - and in the interpretation of their results:

*   how to define the population of interest, and to recruit a representative sample;

*   how the results depend on the way the "skills" of interest are described (e.g. by the researcher, or by the respondents);

*   how to decide whose account of "needs" to accept, especially in the case of conflicts (e.g. between managers and employees).

3. ANTHROPOLOGICAL STUDIES

The importance of the context of use of numerical skills has recently been discussed widely in mathematics education circles. For example, Maier (1980) distinguishes 'folk maths' from 'school maths': 'folk maths' is the maths "that folks do", it "consists of a wide and probably infinite variety of problem-solving strategies and computations that people use" (p. 21).

In recent years, there have been detailed studies of numerate strategies used in work and everyday life, by milk-round dispatchers (Scribner, 1985), by street vendors (Carraher et al., 1985), by lottery bookies (Acioly and Schliemann, 1986) in Brazil, and by shoppers in supermarkets (Lave et al., 1984). These studies document the **distinctive** character of the folk maths **strategies** used in work and everyday settings, as compared with school maths. And they show that these folk strategies are highly **successful**, in context. However, the use of folk strategies **outside** the specific situations where they normally occur is problematic: thus, for example, street vendors who successfully perform many relevant calculations daily in their heads find "similar" calculations to be performed with pencil and paper, outside the context of the market, exceedingly difficult, and make many more errors (Carraher et al., 1985).

In broad terms, there are at least three dimensions of differences pointed to by these discussions. In terms of **computation technology**, school maths largely uses pencil and paper, whereas both folk mathematicians rely on mental computations and algorithms convenient for such use (or calculators when the computations become too difficult).

In terms of **problem-solving strategies** required, school maths very often present pre-formulated problems, with the requisite data provided. Folk problems are seldom clearly defined, and the necessary information must be actively sought (p. 22).

Further, in terms of **level of precision**, school maths normally presents exact numbers and expects (often inexplicably) precise calculations. Folk mathematicians need to be able to cope with estimation in measuring (and sometimes measurement error) and approximation in computation - that is, to make decisions on how much precision is appropriate for their purposes.

Thus analyzing these distinctions, we can use the idea of "folk maths" to produce a richer notion of numeracy, and to begin to describe important dimensions of the context in which it is used.

When problem-solving strategies are used by folk mathematicians, they have several aspects:

(1)   using appropiate **calculations** ;
(2)   **formulating** the problem, usually involving some quantitative modelling ;
(3)   **producing** data or information;
(4)   **analyzing** data;
(5)   **estimating** and **approximating**;
(6)   evaluation, e.g. what degree of precision and what **differences** are worth bothering about (for my present purposes?)

We can see that the aspects of numerate strategies above, notably (2) problem-formulation and modelling, (3) data production, and (4) data analysis, involve a great deal of "statistics".

Further, these aspects of problem-solving strategies are similar to those skills indicated as important for a specialized course in statistics by Anderson (1985, p. 19), and they also recall the aspects of the statistical problem-solving process highlighted in the Open University's service/general course MDST242 (Open University, 1983).

4. CONCERNS IN THE COMMUNITY

Consider the following issues of concern to different groups of people in the institution where I work:

(a)   How can we demonstrate a greater need for books for students of the Social Science Faculty? This question was important to the staff of the Library, as well as to teaching staff and students.

(b)   How can we show that a student counsellor is needed on a particular site of the Polytechnic? This problem was important to students and to teaching staff.

(c)   How much is the bar used by people after 9:00pm - and does this justify the site's remaining open? This question was posed by the caretakers.

(d)   What is the level of use of teaching rooms throughout the week - and does this suggest the need for a broader range of timetabled hours?

A similar list of questions could be produced for a network of community groups. At a Conference in Feb. 1986, the following needs for information and statistics were indicated by various groups:

*   for counter- information to that underlying Local Authority housing policy, by a campaign for secure accommodation for single people - done by simple monitoring of clients visiting the group's office;

*   to show a need for greater resources, by an educational foundation for educationally-needy adolescents;

*   to investigate changes in women's employment patterns, and reasons, by a women's support unit for employment (Local Authority funded);

*   to study the evolving "psychological identity" of ethnic group youth, by a Turkish community group - using a questionnaire survey conducted by a university student on placement.

(For further details, see Cooper, 1986).

Or again, consider a set of examples of "problemising reality" (Freire, 1970) in post-colonial Mozambique (Gerdes, 1985).

These are the sorts of questions that are important to folk mathematicians/ statisticians, and they require the ability to use strategies of producing data, or at least seeking out and using available information, as well as analyzing it. They also require an understanding of the context in which the data are produced - and used.

5. INTRODUCING THE 'BAREFOOT STATISTICIAN'

In searching for a metaphor to capture the sort of statistical numeracy needed by adults in their everyday and working lives, and keeping in mind our earlier discussions of "folk mathematicians", I have been impressed by the idea of a 'barefoot doctor'.

Barefoot doctors or village health workers were first trained in China in the 1930s, especially in Communist-controlled areas of the country (Hillier and Jewell, 1983). Since then, there have been attempts to spread the idea to other parts of Asia (e.g. India and Iran), African (e.g. Tanzania) and Latin America (e.g. Peru and Colombia) (see e.g. Harrison, 1977). We can broadly summarize the characteristics of the work of "idealtypical" barefoot doctors as follows (based on Doyal, 1979, pp. 288-290; and Harrison, 1977).

1. The medical care they provide is based on disease prevention and health education (about e.g. nutrition and sanitation), varying according to the expressed needs of the local community.

1a. They also provide first aid, midwifery, vaccination and taking of samples for lab analysis.

1b. In some settings, they also undertake to diagnose and to treat the most common complaints (affecting the majority of cases of disease); in any case, "the medical help they offer is simplified both in its conceptual framework and in its armoury of treatments", for example, in Peru, it is based on a rigorously simplified manual and a kit of basic medicines (Harrison, 1977, pp. 411-412).

2. They are (often) selected by their local community, on the basis of being trusted local residents, with minimal formal schooling requirements (except for literacy).

3. Training depends, of course, on the breadth of roles to be undertaken (see above), but in any case normally takes place in the rural community itself.

4. The barefoot doctors in some places continue with their normal occupation, receiving no special payment for the medical help they offer.

This account of the barefoot doctor suggests parallel aspects of what might be called a "barefoot statistician's" role:

1. He/she must be able to communicate with their community or 'constituency' (e.g. groups within an educational institution, pressure groups in the neighbourhood, work groups in an underdeveloped country) and to take a lead in responding to the community's expressed needs for information or evidence.

1a. They need to be able to report the results of such searches or investigations in terms that are clear and comprehensible to their constituency, for example making maximum use of graphical methods (see below); they can ensure that results are reported in substantive terms, rather than using "statistical metrics" such as significance levels and r-squared statistics (for further discussion of this point, see Evans, 1982, pp. 247-248); they can ensure that statistical models are used only when they help to communicate trends or differences in the information that are likely to be practically significant to the community concerned.

2. People should normally be (self-)selected into training as barefoot statisticians with only minimal qualifications in "nonstatistical numeracy": i.e. competence in the six basic arithmetic operations (including taking powers and roots) with an appropriate technology and in substituting numbers in an algebraic formula; plus several crucial attitudes of mind: recognition of numbers as essentially a cultural product; not being overwhelmed by the sight or mention of a number of algebraic symbol; acceptance of the appropriateness and the limitations of quantiative arguments. Clearly these numerate skills will themselves be enhanced by the person's involvement in the training.

3. During training, the links of the barefoot statistician with their communities should be maintained by: encouraging students to bring problems from their daily activities; showing the importance of the context and methods of data production for the interpretation of the results; using teaching methods which encourage the active use of investigations and projects, preferably on issues which can be fed back into the community.

4. The barefoot statistician is, of course, in no way a professional statistician, but should have links with a "consultant statistician" who can offer - advice - and - competence-building - in connection with statistical approaches which fall outside of the barefoot statistician's training, but which may appear to be useful for investigating a specific issue; for example, the use of modelling procedures.

6. CONTENT OF A BAREFOOT STATISTICIAN'S TRAINING

It would be tempting to set down a list of topics which the training of a barefoot statistician should include, but that would be ignoring earlier points made here about the necessary responsiveness of the barefoot statistician towards his/her community, and the relationship between the barefoot statistician and his/her knowledge (cf. Mellin-Olsen, 1984). Note also that no assumptions about the setting of the training are made (but see the discussion of "Resources" in the next section). However, the following issues can be highlighted:

(a)   The **production of data** should be emphasized, not only the techniques available, but also the way the process is shaped by decisions which depend on the commitments of the barefoot statistician and the community, and the constraints of the context. For example, decisions about both sampling method and sample size depend on the resources available.

(b)   The barefoot statistician needs knowledge of, and access to, government (national and local), community and institutional **sources** of relevant information. For example; ndicators to do with the local economy, and local health and educational levels.

(c)   Whether presenting the results from primary data, or representing information from secondary sources, the barefoot statistician needs to be conversant with a variety of methods for the **presentation of data** to particular audiences; there are many resources available here, ranging from Exploratory Data Analysis (e.g. Open University, 1983) to more classical presentations (e.g. Croxton, Cowden and Klein, 1939/1955/1967; and the Vienna Method, ISI, 1985).

(d)   The ability of the barefoot statistician to **estimate and approximate** needs to be supported and developed.

(e)   The usefulness of **algorithms**, as an aid to decision-making for the barefoot statistician, should be investigated; for example, the specifying of levels of measurement of relevant variables (e.g. Hoyer, 1986) can be a helpful way to determine which summary statistics to use (see, for example, the inside front cover of Blalock, 1972).There may also be a role for "**expert systems**" (Hand,1985).

(f)   The usefulness of computers or other **computational devices** will depend to an extent on the resources available to particular barefoot statisticians; however, it must be recognised that, for many applications, the speed and power (in calculations and in communications) of the computer is making it effectively indispensable.

(g)   Wherever possible, an effort should be made to build on the **intuitive conceptions** (e.g. of 'probability') and intuitive methods of barefoot statisticians; to take a provocative example: if it were necessary to show how compound probabilities should be calculated, could we not use experiences such as those of bookies who use memorized knowledge or prepared tables of permutations and combinations in taking bets for a lottery game, rather than starting our teaching from scratch (cf. Acioly and Schliemann, 1986).(NOTE 1.)

(h)   As both a teaching and an assessment method (as necessary), **investigations and projects** can be central to the training.

7. RESOURCES FOR BAREFOOT STATISTICIANS

The following give examples of some of the variety of resources available to barefoot statisticians, potential and practising, in the locale where I work and live, namely North London in the U.K.

First, some courses and course materials.

MDST242 (Open University, 1983): This (1/2 credit) course requires 160 hours approximately. In its construction, the topic areas of application - namely economics, education and health - were considered to be at least as important as the "statistical thread" - statistical concepts and techniques. It is available as an institutionally-paced course during the O.U. academic year, and has been exceedingly successful, using the indicator of pass rate. The materials are accessible nationally in some public libraries and bookshops.

QS140:Learning to Work with Data (Middlesex Polytechnic,1984): This course (one semester, 180 hours approx., including course contact) can be followed with no pre-requisites within the flexible Modular Degree Scheme - which offers variable course loads and a Summer School ,with part-time students easily catered for. It concentrates on data sources, graphical and tabular presentation, EDA, simple line-fitting (sub-batch median method) and "intuitive" inference, and attempts to exclude techniques where the mechanics of computation "crowd out" understanding of the applications.

POSE (Project on Statistical Education) (Schools Council, 1980): This set of resources includes 27 teaching units, each with its associated set of teachers' notes and pupils' resource sheets. The materials are widely used in schools, and in some institutions of further and higher education, and there is a national network of coordinators of statistics "across the curriculum". The units certainly address problems arising in the immediate context of students' lives, as well as national policy problems, and many, if not all, could be used with aspiring adult barefoot statisticians.

Next, some networks and groups.

Radical Statistics Group : This is a national network of statisticians and social scientists in the U.K. committed to the critique of statistics as used in the policy-making process (see e.g. Irvine Miles and Evans, 1979; and Runnymede Trust and Radical Statistics Race Group, 1980/1987) It is also committed to the competence-building of potential barefoot statisticians; examples of the work of various subgroups are:

*   an explication and critique of three recent pieces of quantitative educational research, written for practising teachers (Radical Statistics Education Group, 1982);

*   a presentation and critique of available statistics on the military balance, the arms race and the effects of nuclear war, mainly for campaigners for nuclear disarmament (Radical Statistics Nuclear Disarmament Group, 1982);

*   a handbook for workers needing to detect occupational health hazards (Radical Statistics Health Group, 1982);

*   booklets aiming to criticize and to rework the figures presented by the U.K. Secretaries of State for Health, and for Education, to support claims that resources available to the National Health Service, and to the educational system, have been growing (Radical Statistics Health Group, 1987; Radical Statistics Education Group, 1987).

Community Research Advisory Centre (Polytechnic of North London, London N.5): This Centre, established within the Survey Research Unit at P.N.L., has carried out a survey of 220 groups in four North London Boroughs to identify the origins, conduct and outcomes of community research. The Centre has recently held a day conference to share problems, methods and findings and to offer competence-building workshops on basic research methods, such as interviewing, question-naire design, statistical processing, information searching, and dissemination of project results. For further information, see Cooper (1986).

Then, there are databases such as the Data Archive at the University of Essex, and opportunities for up-and-coming barefoot statisticians, such as the BBC's Domes-day Project and the University of London Institute of Education's annual Statistics Competition (Hawkins,1986).

CONCLUSIONS

1. In designing courses in mathematics for adults, it may be helpful to think of these adults as **barefoot statisticians,** and of community research as an appropriate "microworld", or context in which to locate as meaningful the contents of the course . This may be useful for various sorts of courses, such as :

   - a 2-year course in maths/stats./research methods in a social science degree;
   - a single maths course in a liberal arts/ "modular" degree scheme;
   - a post-numeracy course in Further Education; or
   - courses at school; e.g. there are many similarities between the course QM130 (see sec. 7 above) and current GCSE courses from the Midland Examining Group and the Northern Examining Authority.

2. The contents of an applied statistics course include a number of issues with potential for leading to topics of interest for, say, mathematical investigations; for example: measurement scales, leading to number systems; frequency distributions and modelling, leading to graphs and functions; distributions and sampling, leading to probability; and hypothesis tests leading to logic and probability .

3. Further, a wide variety of resources exist for training barefoot statisticians outside of conventional taught courses. In evaluating the usefulness of particular resources, criteria of **accessibility and flexibility** for a particular constituency are paramount, since adult students already have a wide range of **commitments** to their families and to their community.

4. One particular need of barefoot statisticians is for access to a "consultant statistician" - and, by the same token, "community consulting" is a potential growth area for statisticians.

5. There is a need in statistics education for a variety of studies of the use of statistics in work and everyday contexts.

REFERENCES

Acioly, N.M. &
Schliemann, A.D.(1986)   *"Intuitive Mathematics and Schooling in Understanding a Lottery Game"*. Paper given at Psychology of Maths Education Conference, City University, London, 20-25 July.

Anderson, Clive (1985)   "Whither the Teaching of Statistics? The Teaching Methods"; in *ASLIP and Centre for Statistical Education*.

Association of Statistics
Lecturers in Polytechnics
(ASLIP) and Centre for
Statistical
Education   (1985)   *Whither Statistical Education?* Proceedings of a one-day workshop, Centre for Statistical Education, Sheffield.

Bailey, D.E.   (1981)   *Mathematics in Employment (16-18)*: Report, Feb. University of Bath.

Blalock, H.M.   (1972)   *Social Statistics*, 2nd ed. McGraw-Hill.

Carraher, D.
Carraher, T. &
Schliemann, A. (1985)   "Mathematics in the Streets and in Schools". *British Journal of Developmental Psychology, 3, 21-29.*

Cockcroft
Committee   (1982)   *Mathematics Counts*. H.M.S.0.

Cooper,
Libby   (1986)   *"Is there a Case for Community-based research?"*. Report
of a Conference held at Polytechnic of North London
in Feb. 1986. Community Research Advisory Centre,
PNL, London N.5.

Croxton, F.E.,
Cowden, D.J. &
Klein, S.
    (1939/1955/1967)   *Applied General Statistics*. Pitman/Prentice Hall.

Doyal , Lesley   (1979)   *The Political Economy of Health -*
Pluto Press.

Evans, Jeff   (1982)   "After Fifteen Thousand Hours: Where Do We Go
From Here?" - *School Organization* - 2,3, 239-253.

Evans, Jeff   (1988)   "The Politics of Numeracy"in Paul Ernest ed. - *Mathematics Teaching: the State of the Art*. - Falmer Press.

Evans, Jeff   (1990)   *"Mathematics Learning and the Discourse of 'Critical
Citizenship'"* pp 93-95 in R. Noss et al. eds., Political
Dimensions of Mathematics Education: Action and
Critique; Proceedings of the First International Con-
ference (PDME.1). Dept. of MSC, University of Lon-
don institute of Education.

Freire, P.   (1970)   *The Pedagogy of the Oppressed* - Penguin.

Gerdes, P.   (1985)   "Conditions and Strategies for Emancipatory Mathematics Education in Undeveloped Countries"- *For
the Learning of Mathematics* - 5,1 (Feb.), 15-20.

Hand, David J. (1985)   "Choice of Statistical Technique" - *Bulletin of the ISI*

Harris, Mary   (1986)   *"Maths in Work: Skills Frequency Scores"*,
mimeo.(Available from: Maths in Work, ILECC, John
Ruskin St., London SE5 0PQ).

Harrison, Paul (1977)   "Basic Health Delivery in the Third World" - *New
Scientist* - 17 (Feb.), 411-413.

Hawkins, Anne (1986)   *"A National Schools Statistics Competition"* Paper given
at ICOTS II, Univ. of Victoria, B.C. 11-15 Aug.

Hillier, S.M. &
Jewell, J.A.   (1983)   *Health Care and Traditional Medicine in China 1800-1982* - Routledge and Kegan Paul.

Holmes, Peter   (1985)   *Statistical Needs of Non-Specialist Young Workers:* A Report on a Survey Carried Out for the Statistical Education Group (16-19), Centre for Statistical Education, Sheffield.

Hoyer, B   (1986)   *"What Every Statistician Should Know - But Probably Doesn't - About Levels of Measurements"*. Paper given at ICOTS II, University of Victoria, B.C., 11-15 August.

Irvine, John,
Miles, Ian &
Evans, Jeff, eds. (1979)   *Demystifying Social Statistics - .* Pluto Press.

ISI   (1985)   *"The Vienna Method of Statistical Presentation"*. Handout for exhibition at ISI, Amsterdam.

Lave, Jean,
Murtaugh, M. &
de la Rocha, O. (1984)   *"The Dialectic of Arithmetic in Grocery Shopping"*, in Lave and Rogoff, eds.

Lave, J. and
Rogoff, B., eds. (1984)   *Everyday Cognition: Its Development in Social Context.* Harvard University Press.

Maier, E.   (1980)   "Folk Mathematics". *Mathematics Teaching, No. 93*, Dec.

Mellin-Olsen,
Stieg   (1984)   *"The Politicization of Mathematics Education"*. Lecture at the University of London, Institute of Education, 28 November; mimeo, Bergen College of Education, Norway.

Middlesex
Polytechnic   (1984)   *"QM130: Learning to work with Data"*.(Contact: Ivan Rappaport, Middlesex Polytechnic, London N.14, U.K.)

Open University.(1983)   MDST242: *Statistics in Society.* Open University Press.

215

Radical Statistics
Education
Group   (1982)   *Reading Between the Numbers:* A Critical Guide to
                                 Educational Research. (c/o London Hazards Centre,
                                 3rd Floor Headland Ho., 308 Gray's Inn Rd., London
                                 WC1X 8DS.)

Radical Statistics
Education
Group   (1987)   Figuring out educational spending. *Radical Statistics,*
                                 c/o London Hazards Centre.

Radical Statistics
Health Group   (1982)   "Two Statistical Methods for Detecting Health Haz-
                                 ards at Work". *Radical Statistics* , c/o London Hazards
                                 Centre.

Radical Statistics
Health Group   (1987)   Facing the Figures : what really is happening in the
                                 National Health Service - *Radical Statistics*, c/o Lon-
                                 don Hazards Centre.

Radical Statistics
Nuclear Disarmament
Group   (1982)   The Nuclear Numbers Game: Understanding the Sta-
                                 tistics Behind the Bombs. *Radical Statistics*, c/o Lon-
                                 don Hazards Centre.

Radical Statistics Race
Group   (1980 / 1987)   *Britain's Black Population*, Gower.

Schools Council (1980)   *Teaching Statistics 11-16*, Foulsham Educational.

Scribner, S.   (1984)   *"Studying Working Intelligence"*, in Lave and Rogoff,
                                 eds.

Sewell, Brigid   (1981)   *Use of Mathematics by Adults in Everyday Life.* A.C.A.C.E.

Smith A.   (1973).   Generic Skills for Occupational Training - Prince Albert,
                                 Sask. Saskatchewan Newstart, Inc. (available as ERIC
                                 ED 083385).

Wolf, Alison   (1984)   *Practical Mathematics at Work - Learning through YTS -*
                                 Univ. of London Institute of Education,
                                 Research Paper

LIST OF CONTRIBUTORS

John Abraham is a researcher in the sociology of education at the University of Sussex.

Neil Bibby is lecturer in mathematics education at the University of Exeter.

Barry Cooper is lecturer in education at the University of Sussex.

Paul Ernest is senior lecturer in mathematics education at the University of Exeter.

Jeff Evans is senior lecturer in statistics at Middlesex Polytechnic.

Zelda Isaacson was senior lecturer in mathematics education at the Polytechnic of North London, and is now a part-time lecturer, researcher and consultant.

Stephen Lerman is senior lecturer in mathematics education at South Bank Polytechnic, and co-convenor of the group for Research into Social Perspectives of Mathematics Education.

Marilyn Nickson was principal lecturer in mathematics education at Anglia Polytechnic. She is now a writer and Research Officer at the University of Cambridge Local Examinations Syndicate and co-convenor of the group for Research into Social Perspectives of Mathematics Education.

Richard Noss is senior lecturer in mathematics education at the Institute of Education, University of London.

David Pimm is senior lecturer in mathematics education at the Open University.

Leo Rogers is senior lecturer in mathematics education at Roehampton Institute.

Rosalinde Scott-Hodgetts is principal lecturer in mathematics education at South Bank Polytechnic.

Richard Winter is a sociologist of education and a professor at Anglia Polytechnic University.